W9-BVR-556

Digital Filters

PRENTICE-HALL SIGNAL PROCESSING SERIES
Alan V. Oppenheimer, Editor

Andrews and Hunt   *Digital Image Restoration*
Brigham   *The Fast Fourier Transform*
Hamming   *Digital Filters*
Oppenheim and Schafer   *Digital Signal Processing*
Rabiner and Gold   *Theory and Applications of Digital Signal Processing*

DIGITAL FILTERS

R. W. HAMMING

Bell Laboratories
and
Naval Postgraduate School

Prentice-Hall, Inc., Englewood Cliffs, New Jersey 07632

Library of Congress Cataloging in Publication Data

HAMMING, RICHARD WESLEY, 1915–
    Digital filters.

    Bibliography: p. 219
    1. Digital filters (Mathematics) I. Title.
QA297.H26   515'.23   76–30514
ISBN 0–13–212571–4

© 1977 by Prentice-Hall, Inc.
Englewood Cliffs, N.J. 07632

All rights reserved. No part of this book
may be reproduced in any form or by any means
without permission in writing from the publisher.

Printed in the United States of America

10 9 8 7 6 5 4 3 2 1

PRENTICE-HALL INTERNATIONAL, INC., *London*
PRENTICE-HALL OF AUSTRALIA, PTY. LTD., *Sydney*
PRENTICE-HALL OF CANADA, LTD., *Toronto*
PRENTICE-HALL OF INDIA PRIVATE LIMITED, *New Delhi*
PRENTICE-HALL OF JAPAN, INC., *Tokyo*
PRENTICE-HALL OF SOUTHEAST ASIA PTE. LTD., *Singapore*
WHITEHALL BOOKS LIMITED, *Wellington, New Zealand*

Contents

v

10. The Finite Fourier Series   *169*

11. Recursive Filters   *180*

12. Chebyshev Approximation and Chebyshev Filters   *196*

13. **Some Practical Details**   **213**

Preface

This book presents an introduction to digital filters that is relatively free from jargon, that does not depend on special prior knowledge of electrical engineering, and that at the same time covers the fundamentals. The ideas, methods, and results in the field of digital filters are related to

1. statistics, especially time series analysis,
2. numerical analysis,
3. analog filters,
4. sampled data control systems,
5. econometrics,
6. electrical engineering, and
7. digital signal processing.

The processes of smoothing, predicting, differentiating, integrating, separating signals (filtering), and removing the noise of measurement are very common operations. Often such processes are linear transformations on the data and so are digital filters. Numerous interconnections between the fields and processes are discussed in the book.

Conversely, all linear operations on some data are equivalent to filtering, since a filter is any arbitrary linear operation on the data. Thus when processing data, people are constantly filtering without knowing that they are doing it; often they (a) remove, (b) insert, (c) misidentify, or (d) confuse various effects in the original system being studied with effects due to the data processing method used. As a result, they confuse what the data is trying to say with what they have done to the data. On the other hand, they are also ignorant of the possibilities and the power of simple linear operations applied to their data, let alone which ones to choose. The widespread availability of minicomputers has greatly increased the need to understand the field of digital filters, for these computers are often used to gather and process experimental data.

Because digital filters are of importance to a wide variety of people, the material is developed with as few assumptions as possible about the background knowledge necessary. In particular, we assume a knowledge of elementary calculus, a little differential equations, a smattering of numerical analysis, a little statistics (which we will review in Sections 1.5 and 1.6), plus some general scientific sophistication. Again, we do *not* assume an extensive electrical engineering background, as is so often true in texts on digital filters.

Finally, we are interested in presenting the *ideas* of the field and will not always give the "best" methods for designing very complex filters; in an introductory course we are content to offer elementary, broadly applicable methods for designing simple filters. Because it is not an advanced text on digital filters, many references to advanced texts and articles are not appropriate. Instead we refer the reader to a few standard books in which additional information can be found. The references are indicated in the text by [L, p.], giving the book label, L, and the page(s). They appear at the end of the book.

Much repetition exists in the material. Experience shows that the reader often becomes so involved with the details of designing filters or with mathematical theory that he loses sight of where and how the topic being studied fits into the whole plan. Filters are designed ultimately to process data, but again experience shows that the display of large segments of data, before and after processing, communicates very little to the beginner. Thus such plots are rarely given.

As always an author is deeply indebted to others—in this case, to his many colleagues at the Bell Laboratories. Special mention goes to J. W. Tukey, who first introduced him to the field, and to J. F. Kaiser, who taught him much of what is presented here. Thanks are due Ron Crochiere, Roger Pinkham, and others for reading and commenting on the notes for the book. Thanks are also due B. W. Kernighan for much help and Esther Coke for testing the material for readability. No less important are the many secretaries who have helped, from Geri Marky to Carmela Scrocca. Barbara B. Bottini prepared the drawings. However, the author alone is responsible for the defects in the book, for the ultimate choice of what was done was his.

R. W. HAMMING

Digital Filters

1

Introduction

1.1 WHAT IS A DIGITAL FILTER?

Many times in our current society we measure a variable quantity. Some examples include blood pressure, earthquake displacements, voltage from a voice signal of a telephone conversation, brightness of a variable star, population of a city, waves falling on a beach, and the probability of death. All these measurements vary with time, and we regard them as functions of time— $x(t)$ in mathematical notation. And we may be concerned with blood pressure measurements from moment to moment or from year to year. Furthermore, we may be concerned with functions whose independent variable is not time, such as the number of nuclear particles in a given experiment as a function of their energy.

Usually these variables can be regarded as varying continuously (analog signals) even if, as with the population of a city or a bacterial colony, the number being measured must change by unit amounts.

For technical reasons, instead of recording the signal $x(t)$, we often merely record equally spaced samples x_n of the function $x(t)$. The famous *sampling theorem*, which will be discussed in Chapter 8, gives the conditions on the signal that justify this sampling process. Moreover, when the samples are taken, they are not recorded with infinite precision but are rounded off (sometimes chopped off) to comparatively few digits. This procedure is often called *quantizing* the samples. It is these quantized samples that are available for the processing that we do. We do the processing in order to understand what the function reveals about the underlying phenomena that gave rise to the observations, and digital filters are the main processing tool.

Suppose that the sequence of numbers x_n is such a set of equally spaced measurements of some quantity $x(t)$, where n is an integer and t is a continuous variable. Typically, t represents time but not necessarily so. We are using the notation $x_n \equiv x(n)$. If the sequence y_n is computed by the formula

$$y_n = \sum_{k=-\infty}^{\infty} c_k x_{n-k} + \sum_{k=1}^{\infty} d_k y_{n-k}$$

then this formula defines a *digital filter*. The coefficients c_k and d_k are constants. Thus a digital filter is merely a linear combination of equally spaced samples x_{n-k} of some function $x(t)$, together with the computed values of the output y_{n-k}. For each successive n, the formula shifts one data point along the string of data points, x_{n-k}.

When, as above, the second summation is restricted to the range of *positive* indices k, the equations are easily solved for the y_n. But there is no theoretical reason for this restriction. If we did not so restrict the index, we would, of course, be driven to solving a system of linear equations for the unknowns y_n. In many situations, the cost of gathering the information is so great that the cost of solving the corresponding high-order set of linear equations is not a serious question (since the equations tend to have large terms along the main diagonal). However, except for Section 11.6, we will not pursue this point further.

Various special cases of this formula occur frequently and should be familiar to most readers. Indeed, such formulas are so commonplace that a book could be devoted to listing them. In the case where all the coefficients d_k of the y_{n-k} are zero, the filter is called *nonrecursive*; otherwise it is a *recursive* filter. A familiar example of a nonrecursive filter is the widely used smoothing by 5s (Section 3.2)

$$y_n = \frac{1}{5}[x_{n-2} + x_{n-1} + x_n + x_{n+1} + x_{n+2}]$$

Another example is the least-squares smoothing formula derived by passing a least-squares cubic through five equally spaced values x_k and then using the value of the cubic at the midpoint as the smoothed value. The formula for this smoothed value (which will be derived in Section 3.3) is

$$y_n = \frac{1}{35}[-3x_{n-2} + 12x_{n-1} + 17x_n + 12x_{n+1} - 3x_{n+2}]$$

Many formulas for predicting stock market prices, as well as other values of a time series, also fall in the form of a nonrecursive filter.

On the other hand, an example of a *recursive* filter is the trapezoid formula for numerical integration (Section 3.4)

$$y_{n+1} = y_n + \frac{1}{2}[x_n + x_{n+1}]$$

It is immediately obvious that a recursive formula can, as it were, remember all the past data, since the y_n value on the right-hand side of the equation enters into the computation of y_{n+1} and hence into y_{n+2}, and so on.

In some situations, the values of x_k and y_k for $k < 0$ are not available. Formulas which do not use these values are called *physically realizable*. This name is misleading and should probably be avoided in discussions because physically nonrealizable filters can be programmed on digital computers when, as is typically true, all the data is available *before* the computation starts. However, when predicting *future* values from some data, a physically realizable filter must be used. Physically realizable filters are also called *causal*, for if time is the independent variable, then they do not react to future events but only to past ones (causes).

For practical reasons, it is necessary to consider only *finite-length* digital filters, meaning that only a finite number of terms occur in the filter and that all the rest of the coefficients are zero. For instance, if all the d_k are zero and only a finite number of the c_k are not zero, then we have a finite-length non-recursive filter. Even with such a filter it is immediately evident that for a finite length of data we will not be able to compute values near the ends of the data. Thus there is a loss of data in the sense that there is less output than there is input. For a long sequence of data points this situation may be un-important; but for a fairly short sequence it is serious and tends to force us to use fairly short-span finite filters. The common practice of supplying a sequence of zero values beyond one or both ends of the data, so that as many output values can be computed as there were original input values, is danger-ous. (See also Section 10.7)

The filter coefficients c_k and d_k are assumed to be *constants* and unvarying with time. Such filters, called *time-invariant filters*, are the ones most com-monly used in practice. Time-varying filters are occasionally useful, but their discussion is beyond the scope of this book.

Finally, it must be realized that the evaluation will be done by using finite-length numbers. The process of *quantizing* the numbers produces roundoff, both in the coefficients of the filter and in the arithmetic done in the machine.

Consequently, there are roundoff errors in the final numbers y_n that are obtained. It is often convenient to *think* in terms of an infinite number of terms in a filter with each term of infinite precision and of using infinite precision arithmetic; in the end, however, we must deal with reality. Furthermore, the way in which we arrange to do the arithmetic can sometimes significantly alter the numbers that result from the computation. We shall consider this topic more closely in Chapters 12 and 13.

1.2 WHY SHOULD WE CARE ABOUT DIGITAL FILTERS?

The word *filter* is derived from electrical engineering, where filters are used to transform electrical signals from one form to another, especially to eliminate (filter out) various frequencies in a signal. As we have already seen, a digital filter is a linear combination of the input data x_n and possibly the output data y_n and includes many of the operations that we do when processing a signal.

For convenience we suppose that we sample at unit time and that we represent the nth sample of the signal as x_n. Thus we may think of blood pressure, a brain wave, the height of a wave on a beach, or a stock market price as a continuous signal that we have sampled (and quantized) at unit times in order to obtain our sequence of data x_n. In the stock market case we often take the integral over a period, say a week, and record only the total amount each week, although the underlying idea of continuous variation of the price or whatever else is being measured (say the the rate at which shares are traded) still exists. Given such a signal, we may want to differentiate, integrate, sum, difference, smooth, extrapolate, analyze for periodicity, or possibly remove the noise; all these and many others, are linear operations. Therefore, in the digital form, the operations are *digital filters*.

Widespread use of minicomputers in science and engineering has greatly increased the number of digital signals recorded and processed. Since we are already processing such data in a linear fashion, it is necessary to understand the alterations and distortions that these filters produce. Moreover, because digital transmission is so much more noise resistant than analog signal transmission, a world dominated by digital transmission is rapidly approaching. Thus again we are impelled to study exactly what digital filters do, or can be designed to do, to various signals.

Occasionally we read about *sampled data systems*. Here the signal is sampled, but the sampled value is *not* quantized. Such systems will not be considered in this book.

Exercises

1.2-1   If the measurements are made at times $t_n = t_0 + n\,\Delta t$ $(n = 0, 1, 2,\ldots)$, write the formula $t'_n = f(t_n)$ that produces unit spacing in t' and starts at $t'_0 = 0$.

1.2-2   List ten sources, other than those in the text, of signals that might be filtered.

1.2-3   Write Simpson's integration formula as a recursive filter.

1.2-4   Compute the first five output values of the filter

$$y_n = ay_{n-1} + x_n$$

for the input $y_0 = 0$, $x_n = 1$ for all n.

1.2-5   Compute the successive values of the filter

$$y_n = ay_{n-1} + x_n$$

where $y_0 = 0$, $x_1 = 1$ and all other $x_n = 0$. Give the formula for y_n.

1.3   HOW SHALL WE TREAT THE SUBJECT?

Much of the theory, both as to design and use of digital filters, originated in the field of analog filters. If one is already familiar with the field, then it is reasonable to build on this knowledge. Today however, the average person who needs to know about digital filters has no such background, and so it is foolish to base the development on the analog approach. Consequently, we assume no such familiarity and merely mention the corresponding jargon when necessary.

The statistics field has also contributed to the theory of digital filters. In particular, the subject of time series is closely related and has contributed its own jargon.

Textbooks in numerical analysis have many formulas that are linear combinations of equally spaced data, and thus such formulas are equivalent to digital filters. Since the elements of numerical analysis are now more widely known than those of other fields of application, we will select many of our examples from numerical analysis.

The fundamental approach common to all the special fields is based on (a) the Fourier series, both discrete and continuous, and (b) the use of the Fourier integral. They are the mathematical tools for understanding and manipulating linear formulas, and we must take the time to develop them, for they are rarely taught outside of electrical engineering courses these days. However, we will avoid becoming too involved with mathematical rigor, which all too often tends to become rigor mortis. Nor do we develop all the mathematical theory before showing its use; instead we regularly give applications of the theory just covered in order to show both its relevance and its use. In this way, hopefully, much of the mathematics will become more obvious to the nonmathematically inclined.

1.4 GENERAL-PURPOSE VERSUS SPECIAL-PURPOSE COMPUTERS

Digital filtering occurs on both special-purpose and general-purpose digital computers. Even though numerical computations are also done on both types, most introductory textbooks on the subject deal only with computing done on general-purpose computers; similarly most of the discussion is confined to filtering done on general-purpose computers.

This remark should not be interpreted as meaning that the field of special-purpose computers is unimportant. Rather it is an indication that computation on a general-purpose computer is usually much less restrictive. Therefore, in a first presentation, we concentrate on the main ideas while ignoring the details of the particular computer being used. Special-purpose digital computers are rapidly increasing in importance, primarily because of the availability of inexpensive, large-scale integrated circuits, as well as the fact that many of the operations that we wish to perform are sometimes beyond the scope (in either an economic or a time sense, or both) of current (and foreseeable) general-purpose computers.

1.5 ASSUMED STATISTICAL BACKGROUND

A set of measurements is called a *sample*. The word "sample" is used both for an individual measurement and for the set of measurements, even if they are repeated measurements of the same thing. This usage occurs because the statistician is thinking of an underlying *population* or *ensemble* of possible measurements and you have obtained one possible set of results (one realization). He is concerned with the probability of obtaining the particular observed result and with the effects of repetitions of the experiment. The measurements of a sample can all be at a single point. For instance, the sample can be a number of measurements of the length of a particular wire. The measurements can also be scattered at various places in the range of a function—for example, the velocity of a boat at various times of a day.

Often a *model* for the distribution of the measurements must be found; we want to think about the ensemble from which we have drawn the particular sample.

To illustrate, if L is the measured length of the wire just mentioned, then we model these measurements by

$$P\{L \leq x\} = P(x)$$

For $P\{L \leq x\}$ read "probability of L being less than or equal to x." $P(x)$ is thus the probability that the measured length L is less than or equal to x; $P(x)$ is called the *cumulative distribution function for L*. In many situations, $P(x)$ has a derivative $p(x)$; that is,

$$\frac{dP}{dx} = p(x) \quad \text{and} \quad P\{a < L \leq b\} = \int_a^b p(x)\,dx$$

Then $p(x)$ is called either the *density* or the *probability density for L*.

A common density that occurs in such situations as measuring the length of a piece of wire is the gaussian (or normal) distribution

$$p(x) = \frac{1}{\sigma\sqrt{2\pi}} e^{-(x-\mu)^2/2\sigma^2} \qquad (-\infty < x < \infty)$$

where μ and σ are parameters whose values depend on the particular situation being modeled. See Fig. 1.5-1(b).

Another example of a model occurs in roundoff theory. It is reasonable to suppose that the roundoff error made when a number is quantized (rounded off) is "uniformly distributed" from $-\frac{1}{2}$ to $\frac{1}{2}$ in the last digit kept. Therefore

$$p(x) \equiv \begin{cases} 1, & -\frac{1}{2} \leq x \leq \frac{1}{2} \\ 0, & |x| > \frac{1}{2} \end{cases}$$

See Fig. 1.5-1(a).

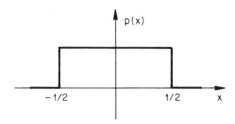

(a) ROUNDOFF DISTRIBUTION

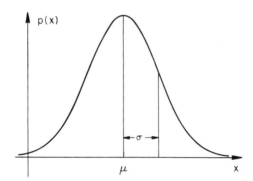

(b) GAUSSIAN DISTRIBUTION

$$p(x) = \frac{1}{\sigma\sqrt{2\pi}} e^{-\frac{(x-\mu)^2}{2\sigma^2}}$$

FIGURE 1.5-1

A commonly computed characteristic of a random quantity such as L or, alternatively, of a density $p(x)$ is the *average* or *expected value* (also called *mean value*). It is denoted Ave (L) or $E(L)$ and is defined by

$$\text{Ave}\,(L) \equiv E(L) \equiv \int_{-\infty}^{\infty} x p(x)\,dx$$

For the roundoff example,

$$\int_{-\infty}^{\infty} x p(x)\,dx = \int_{-1/2}^{1/2} x\,dx = 0$$

For the gaussian example,

$$E(L) = \frac{1}{\sigma\sqrt{2\pi}} \int_{-\infty}^{\infty} x e^{-1/2[(x-\mu)/\sigma]^2}\, dx$$

Replacing $(x - \mu)/\sigma$ by y, we have

$$E(L) = \sigma \int_{-\infty}^{\infty} y e^{-(1/2)y^2} \frac{dy}{\sqrt{2\pi}} + \mu \int_{-\infty}^{\infty} e^{-(1/2)y^2} \frac{dy}{\sqrt{2\pi}}$$

$$= 0 + \mu = \mu$$

The last equation is true because the first integrand is odd and the second integral gives $P\{-\infty < L < \infty\} = 1$.

We can think of the expectation as an operator $E(\ )$ operating on a function. A moment's thought and it is obvious that the expected value of a constant is the same constant,

$$E(a) = a$$

Again, if x is the variable of the model and a is a constant, then

$$E(ax) = aE(x)$$

and if b is also a constant, then

$$E(ax + b) = aE(x) + b$$

Other "typical values" besides the average are widely used. One is the *mode*, the most frequent value or the one with maximum density. Another is the *median*, the value exceeded by half the distribution. We will not use either of them in this book.

Another commonly computed characteristic of a random quantity or, alternatively, of its distribution is the *variance*. It is denoted Var (L), if L is the random quantity, and defined by

$$\text{Var } (L) = \int_{-\infty}^{\infty} (x - \mu)^2 p(x)\, dx$$

where L has density $p(x)$ and $\mu = E(L)$. It is also denoted by the symbol σ^2.

For the roundoff case, we have

$$\sigma^2 = \int_{-\infty}^{\infty} (x - 0)^2 p(x)\, dx = \int_{-1/2}^{1/2} x^2\, dx = \frac{1}{12}$$

and for the gaussian,

$$\sigma^2 = \text{Var}\,(L) = \frac{1}{\sigma\sqrt{2\pi}} \int_{-\infty}^{\infty} (x - \mu)^2 e^{-1/2[(x-\mu)/\sigma]^2}\, dx$$

If we set

$$x - \mu = t(\sigma\sqrt{2})$$

we obtain

$$\text{Var}\,(x) = \frac{2\sigma^2}{\sqrt{\pi}} \int_{-\infty}^{\infty} t^2 e^{-t^2}\, dt$$

Integration by parts using

$$\begin{cases} te^{-t^2}\, dt = dV, & V = \dfrac{e^{-t^2}}{-2} \\[2mm] U = t, & dU = dt \end{cases}$$

gives

$$\text{Var}\,(x) = \frac{2\sigma^2}{\sqrt{\pi}} \left[-\frac{e^{-t^2}}{2} t \Big|_{-\infty}^{\infty} + \frac{1}{2} \int_{-\infty}^{\infty} e^{-t^2}\, dt \right]$$

$$= \frac{2\sigma^2}{\sqrt{\pi}} \frac{\sqrt{\pi}}{2} = \sigma^2$$

This result shows why we adopted the peculiar form for writing the gaussian distribution: σ is the variance and μ is the average of the gaussian (normal) distribution,

$$p(x) = \frac{1}{\sigma\sqrt{2\pi}} e^{(x-\mu)^2/2\sigma^2} \equiv N(\mu, \sigma^2)$$

It is clear that the variance, which is the sum of the squares of the deviations of the distribution from its average value [weighted by the probability $p(x)$ of occurring], is closely related to the principle of least squares (which states that the "best fit" occurs when the sum of the squares of the errors is minimum). In both cases, it is the sum of the squares of the differences that is used. For the variance, it is the difference from the mean that is used; for a least squares fit, it is the difference of the data from the approximate fit that is used.

Exercises

1.5-1 If the distribution for $p(x)$ is

$$p(x) = \begin{cases} 0, & x < 0 \\ ae^{-ax}, & x \geq 0 \end{cases} \quad (a > 0)$$

show that $\mu = 1/a$ and $\sigma^2 = 1/a^2$.

1.5-2 If the distribution for $p(x)$ is

$$p(x) = \begin{cases} 1 - \dfrac{x}{2}, & 0 \leq x \leq 2 \\ 0, & \text{otherwise} \end{cases}$$

show that $\mu = 2/3$, $\sigma^2 = 2/9$.

1.5-3 Find the mean and variance of

$$p(x) = \begin{cases} \cos 2x, & -\dfrac{\pi}{4} \leq x \leq \dfrac{\pi}{4} \\ 0, & \text{otherwise} \end{cases}$$

1.5-4 For a well-balanced die (singular of dice), calculate the mean and variance of the value on the top face after a "random toss." Do the same for a pair of dice.

1.6 THE DISTRIBUTION OF A STATISTIC

We now turn to what is probably the hardest concept for the beginner in statistics to master, the idea of the distribution of a statistic (such as the mean or the variance of a sample).

Suppose that we have made a set of measurements and that from the sample we have computed one or more statistics. For instance, we may have selected 1000 Americans at random from the entire population of around 200 million and measured their heights. From these heights we can compute the *sample average* $\bar{x}$. The *variance* s^2 *of the sample* is defined by

$$s^2 \equiv \frac{1}{n-1} \sum_{i=1}^{n} (x_i - \bar{x})^2$$

For clarity, it is customary to use Greek letters for the statistics of the model and Latin letters for the corresponding statistics of the sample.

It is good to know these two numbers for the sample that we drew, but if we are to make much use of them, the question immediately arises: If we repeat the whole process again, using a different random sample of 1000 Americans, what could we reasonably expect to get for the average? In short, what is the distribution of the statistic called the "average"? Clearly, repetitions of the whole process of selecting the people, making the measurements, and computing the average will give us a distribution of values for the average $\bar{x}$ (and a distribution for the variance s^2).

In the roundoff example we had a unique model for the basic population from which the roundoffs were drawn, but in the gaussian example we must estimate the two unknown parameters of the population distribution μ and σ^2 from the sample statistics $\bar{x}$ and s^2. We can ask what relation these two sets of numbers have to each other. In textbooks on statistics it is proved that, for any distribution, the average of the sample is an unbiased estimate of the original population average. Similarly, the sample variance s^2 is an unbiased estimate of σ^2/n. Unbiased means that, on the average, your estimates are neither too high nor too low. (That is, the average of the statistic equals the value being estimated.)

TABLE 1.6-1   Relation of sample to population statistics

	Sample	Population
Ave $(x) = \dfrac{1}{n} \sum_{i=1}^{n} x_i$		μ
$s^2 = \dfrac{1}{n-1} \sum_{i=1}^{n} (x_i - \bar{x})^2$		$\dfrac{\sigma^2}{n}$

If the sample is at all large $(n \geq 10)$, then the central limit theorem shows that the statistic called the average has a distribution that is very close to a gaussian (normal) distribution

$$p(x) = \frac{\sqrt{n}}{\sigma\sqrt{2\pi}} e^{-(x-\bar{x})^2 n/2\sigma^2}$$

with parameters $\bar{x}$ and σ^2/n.

Exercise

1.6-1   For the set of measurements 10, 11, 10, 12, 9, 10, 7, 10, 10, 9, compute the mean and variance of the sample and estimate the corresponding population parameters.

1.7 NOISE AMPLIFICATION IN A FILTER

Suppose that we make some measurements. Let x_n be a "true" measurement with an added noise ϵ_n whose expected value $E(\epsilon_n) = 0$. Thus what we record is $x_n + \epsilon_n$. Furthermore, let this noise ϵ_n have a variance of σ^2. What is the corresponding noise in the output of a nonrecursive filter (assuming that the arithmetic we do does not increase the noise)? We also suppose that for the actual measurements, $x_n + \epsilon_n$, the errors ϵ_n are *uncorrelated*. Thus we are assuming that

$$E\{\epsilon_n \epsilon_m\} = \begin{cases} \sigma^2, & m = n \\ 0, & m \neq n \end{cases}$$

The condition of a zero mean is

$$E\{\epsilon_n\} = 0$$

and implies that there is no bias in the measurements. The averaging is, in both cases, over the ensemble of the noise ϵ_n.

A nonrecursive filter is defined by the formula

$$y_n = \sum_{k=-K}^{K} c_k(x_{n-k} + \epsilon_{n-k})$$

The expected value is, therefore, since the E operation applies only to the ϵ_n and not to the c_k or the x_{n-k},

$$E\{y_n\} = \sum_{k=-K}^{K} c_k(x_{n-k} + E\{\epsilon_{n-k}\}) = \sum_{k=-K}^{K} c_k x_{n-k}$$

For the variance calculation, we begin with

$$E\left\{\left[\sum_{k=-K}^{K} c_k(x_{n-k} + \epsilon_{n-k}) - E(y_n)\right]^2\right\}$$

But this is

$$E\left\{\sum_{k=-K}^{K} c_k \epsilon_{n-k}\right\}^2 = E\left\{\left[\sum_{k=-K}^{K} c_k \epsilon_{n-k}\right]\left[\sum_{m=-K}^{K} c_m \epsilon_{n-m}\right]\right\}$$

Since $E(\epsilon_n) = 0$ and for $m \neq n$ $E(\epsilon_n \epsilon_m) = 0$, multiplying out and applying the operator E to the ϵ_n leave only the terms

$$\sum c_k^2 E\{\epsilon_k^2\} = \sum c_k^2 \sigma^2 = \sigma^2 \sum_{k=-K}^{K} c_k^2$$

Thus the sum of the squares of the coefficients of a filter measures the noise amplification of the filtering process. It is for this reason that the sum of the squares of the coefficients of a nonrecursive filter plays a significant role in the theory.

Exercises

1.7-1   Apply this last formula, using the roundoff noise model.

1.7-2   What is the noise amplification of the least-squares cubic of Section 1.1?

1.7-3   What is the noise amplification of smoothing by $5s$? Answer: $\sigma^2 = \frac{1}{5}$

1.7-4   Show that the minimum noise amplification of a five-term nonrecursive filter, with the sum of the coefficients equal to 1, is the smoothing by $5s$.

2

The Frequency Approach

2.1 INTRODUCTION

The purpose of this chapter is to show, for linear digital filter design, why and in what sense the use of sines and cosines of the independent variable t is preferable to the classical use of polynomials in t. The approximation of a function by a polynomial is generally emphasized in mathematics, statistics, and numerical analysis. For instance, in Newton's method for finding a zero of a function $g(t)$, the function is locally replaced by the tangent line—a linear equation in t. Again, a Taylor's expansion of a function expresses the function in powers of $t - t_0$. In statistics data is constantly being fitted by polynomials. In the trapazoid rule for integration the function is locally replaced by a straight line. It is natural, therefore, to suppose that in other fields polynomials are the proper functions to use when approximating a given function. Thus we are concerned in this chapter more with the psychological problem of undoing this earlier conditioning in favor of polynomials than with the logical problem of presenting the frequency approach.

We shall show, in three different senses, that the sines and cosines are the proper functions for situations that are relevant to much of data processing on computers. To do so, it is necessary to introduce the concepts of *eigenfunctions* and *eigenvalues* and to show that the concept of the *transfer function* corresponds to the eigenvalues of the process.

Before proceeding, however, we introduce in the next section the most important consequence of sampling a function at equally spaced points. This phenomenon, called *aliasing*, is a common experience for most people; but they are so accustomed to it that they are only vaguely aware of it.

Since the idea of *frequency* is clearly central to the frequency approach, we must be careful to say what it means. Consider, for example, a rectangular wave (or any other shaped wave) that exactly repeats itself 10 times a second; we say that it has a *period* (cycle) of $T = \frac{1}{10}$ second and a *rotational frequency* of 10 hertz (cycles per second). Hertz is abbreviated as Hz when used as a unit of measure. By a period of a function, we mean the *shortest* interval for which the function exactly repeats itself, and the *fundamental frequency* is the corresponding frequency.

The period T and the frequency f are reciprocals of each other. The *angular frequency* ω (in radians) is related to the rotational frequency f by

$$\omega = 2\pi f$$

The use of the angular measure ω is natural in calculus situations, and use of the rotational frequency f is natural in applications. This is the way that we will use them.

The adjective *fundamental* is often dropped, but doing so can lead to confusion. For instance, in Section 4.3 we will decompose a rectangular wave into a sum of sines and cosines and then say that the original waveform has high frequencies in it. Thus confusion can arise concerning the frequency of the original waveform and the frequencies of the terms in the decomposition of the wave into a set of sinusoidal (periodic) functions.

2.2 ALIASING

The phenomenon of *aliasing*, which is basic to sampling data at equally spaced intervals, is not new to the reader who has watched Westerns either on TV or in the movies. As the stagecoach wheels turn faster and faster, they appear to slow down and then to stop. If the increase in speed is great enough, they may seem to go backward, stop, and go forward a number of times. Any actual high rate of rotation of the wheels appears, as a result of the sampling of the pictures, to be "aliased" into a low frequency of rotation. Figure 2.2-1 shows, symbolically, a wheel with four spokes rotating at different rates, and the human mind interprets what is seen as the smallest motion that accounts for the observations.

Another common application of this phenomenon of aliasing due to sampling occurs when a stroboscope is flashed at a rate close to that of a piece

 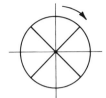 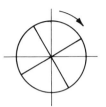

SLOW ROTATION   MIDDLE ROTATION   FASTER ROTATION

APPEARS AS A DIRECT   APPEARS AS IF STAND-   APPEARS AS A BACK-
ROTATION   ING STILL WITH DOUBLE   WARD ROTATION
                       THE NUMBER OF SPOKES

FIGURE 2.2-1

of rotating equipment. If the stroboscope flashes at a rate slightly *less* than the rate of rotation (or some multiple of it), then the flashes make the machine appear to the eye as if it were rotating slowly forward; the closer the rates, the slower the apparent rotation. Again we see that one frequency is aliased into another due to the process of taking equally spaced samples.

In the case of a sinusoid, we are not sampling a rotating wheel but are, in effect, sampling one component, either vertical or horizontal (or in any other direction for that matter). As a result, we can see the phenomenon of aliasing due to sampling at equally intervals in time (the independent variable) as a simple consequence of trigonometric identities.

Consider the sinusoid

$$\cos [2\pi(m + a)t + \phi]$$

where m is an integer, positive or negative, a is the *positive* fractional part of the original rate of rotation, and ϕ is an arbitrary phase angle. Since we are sampling the sinusoid at integer values of t, the reduction of any angle by $2k\pi$ leaves the cosine with the same value. Therefore the sinusoid is equivalent to

$$\cos [2\pi at + \phi]$$

If $a > \frac{1}{2}$, we can remove another 2π, and [since $\cos x = \cos (-x)$]

$$\cos [2\pi(-1 + a)t + \phi] = \cos [2\pi(1 - a)t - \phi]$$

at the sample points. Thus we have shown, using only simple trigonometry, that *at the sample points* any sinusoid of arbitrary frequency is equivalent to a

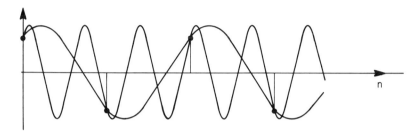

FIGURE 2.2-2   ALIASING

sinusoid with a frequency that lies between 0 and $\frac{1}{2}$, equivalent in the sense that the two sinusoids have the same numerical values at the sample points (see Fig. 2.2-2). In a very real sense, the two frequencies are indistinguishable —a high frequency is aliased into (appears as) a low frequency due solely to the sampling process. Only sinusoids with frequencies low enough so that at least two samples occur in each period are not aliased.

To illustrate, the function

$$y_n = \cos \frac{7\pi}{2} n \qquad (n = 0, \pm 1, \pm 2, \dots)$$

has the same values as

$$y_n = \cos \left(\frac{7\pi}{2} - 4\pi \right) n = \cos \left(\frac{-\pi}{2} n \right) = \cos \frac{\pi}{2} n$$

at all the sample points, and hence the original function is aliased into the lower-frequency function.

Exercises

2.2-1   A machine rotates at 100 Hz. If a strobe light flashes at a rate of 99 per second, what is the apparent motion of the machine? (*Hint:* Take 1/99 second as the unit of time for the sampling.) If the strobe flashes 101 times per second? 98 times?

2.2-2   Find the lowest aliased frequency of $\cos [8\pi n/3 + \pi/3]$. Of $\cos [13\pi n/3 + \pi/3]$.

2.2-3   Repeat the argument for sines in place of cosines. Note carefully the small differences between the cosines and sines.

2.3   THE IDEA OF AN EIGENFUNCTION

The word *eigenfunction* is a half translation from the German of what in the older English texts was called characteristic function, proper function, or natural function.

As a special case of eigenfunctions, consider the multiplication of a square matrix $\mathbf{A} = (a_{i,j})$, of dimension N by N, by a vector $\mathbf{x}$, of dimension N by 1. The product is another vector $\mathbf{y}$, of dimension N by 1,

$$\mathbf{Ax} = \mathbf{y}$$

If $\mathbf{A}$ were the identity matrix, then, of course, the vector $\mathbf{x}$ equals the vector $\mathbf{y}$ in the sense that all the components have the same value. Also, if $\mathbf{x} = 0$, then $\mathbf{y} = 0$, but in the future we shall exclude the function (vector) that is identically zero.

Usually the output vector $\mathbf{y}$ will point in a different direction (in the N-dimensional space) from the input vector $\mathbf{x}$. For the typical matrix $\mathbf{A}$ of dimension N, there will be N different vectors $\mathbf{x}$ such that the corresponding $\mathbf{y}$ will have the *same direction* as did the $\mathbf{x}$, although not necessarily the same length. That is, we will have

$$\mathbf{Ax} = \lambda\mathbf{x}$$

for some constant λ. To see the truth of this remark, we can write the preceding equation in the form

$$(\mathbf{A} - \lambda\mathbf{I})\mathbf{x} = 0$$

where $\mathbf{I}$ is the identity matrix. In order for this equation to have a solution that is not identically zero, it is both necessary and sufficient that the determinant of the system of equations

$$|\mathbf{A} - \lambda\mathbf{I}| = 0$$

This determinant, when expanded, is clearly a polynomial in λ of degree N, and, in general, it will have N distinct zeros $\lambda_1, \lambda_2, \ldots, \lambda_N$, real or complex (to have a multiple zero is a restriction on the matrix $\mathbf{A}$). Thus there are, in general, N distinct λ_k with corresponding vector solutions $\mathbf{x}_k$. (Note that $\mathbf{x}_k$ is a vector and not a component of a vector.) The values λ_k are called the *eigenvalues*, and the $\mathbf{x}_k$ are called the corresponding *eigenvectors*. For a given eigenvalue, the determinant is zero, and the corresponding eigenvector is, of course, determined to within a multiplicative constant.

Why are these eigenvectors important? There are (in general) N distinct eigenvectors, and they can be shown to be *linearly independent*. Therefore they can serve as a basis for representing an arbitrary vector $\mathbf{x}$ of N dimensions. So we can represent an arbitrary vector $\mathbf{x}$ as a linear combination of the N eigenvectors $\mathbf{x}_k$

$$\mathbf{x} = \sum_{k=1}^{N} a_k \mathbf{x}_k$$

If we now multiply this equation on both sides by the matrix $\mathbf{A}$ (technically, we apply the operation $\mathbf{A}$ to the equation), we find that

$$\mathbf{A}\mathbf{x} = \sum_{k=1}^{N} a_k \mathbf{A}\mathbf{x}_k = \sum_{k=1}^{N} a_k \lambda_k \mathbf{x}_k$$

and we see that each eigenvector is multiplied by its corresponding eigenvalue. In the eigenvector representation, the effect of the multiplication by the matrix $\mathbf{A}$ (applying the operation $\mathbf{A}$) is easy to follow. The eigenvectors remain independent of each other.

For instance, the matrix

$$\mathbf{A} = \begin{pmatrix} 1 & 2 \\ 3 & 2 \end{pmatrix}$$

leads to the corresponding determinant

$$|A - \lambda I| = \begin{vmatrix} 1 - \lambda & 2 \\ 3 & 2 - \lambda \end{vmatrix} = 0$$

Upon expansion of the determinant, we get the equation for the eigenvalues

$$\lambda^2 - 3\lambda + 2 - 6 = 0$$

which has zeros $\lambda = 4, -1$. If we use $\lambda_1 = 4$, we obtain for the matrix equation (where we are using the notation $x_{i,j}$ for the jth component of the ith vector $\mathbf{x}_i$)

$$\begin{vmatrix} -3 & 2 \\ 3 & -2 \end{vmatrix} \begin{vmatrix} x_{1,1} \\ x_{1,2} \end{vmatrix} = 0$$

which leads to the single equation

$$-3x_{1,1} + 2x_{1,2} = 0$$

and the corresponding eigenvector

$$\begin{vmatrix} x_{1,1} \\ \dfrac{3x_{1,1}}{2} \end{vmatrix}$$

The value of $x_{1,1}$ is arbitrary, since the rank of the matrix for the eigenvalue is 1. If we use the other eigenvalue $\lambda_2 = -1$, we obtain, correspondingly,

$$\begin{vmatrix} 2 & 2 \\ 3 & 3 \end{vmatrix} \begin{vmatrix} x_{2,1} \\ x_{2,2} \end{vmatrix} = 0$$

with the corresponding eigenvector

$$\begin{vmatrix} x_{2,1} \\ -x_{2,1} \end{vmatrix}$$

These two eigenvectors can represent any arbitrary two-dimensional vector.

Exercises

2.3-1   Find the eigenvalues and eigenvectors of the matrix

$$\mathbf{A} = \begin{vmatrix} 1 & -1 \\ 2 & 0 \end{vmatrix}$$

2.3-2   Given the matrix

$$\mathbf{A} = \begin{vmatrix} 1 & 1 & 1 \\ 1 & 0 & -2 \\ 1 & 0 & 0 \end{vmatrix}$$

find all the eigenvalues and the eigenvector corresponding to the eigenvalue 1.
Answer: $(x_{1,1}, -x_{1,2}, x_{1,3})$

2.4   INVARIANCE UNDER TRANSLATION

In many data processing problems there is no natural origin, and therefore an arbitrary point is selected as the origin (typically, for a time signal, it
is the time when we set $t = 0$, which is arbitrary). From the addition formulas
of trigonometry

$$\sin (x + y) = \sin x \cos y + \cos x \sin y$$
$$\cos (x + y) = \cos x \cos y - \sin x \sin y$$

it is easy to see that when $x = x' + h$,

$$A \sin x + B \cos x$$

will go into

$$A' \sin x' + B' \cos x'$$

where we have

$$A' = A \cos h - B \sin h$$
$$B' = A \sin h + B \cos h$$

Squaring each of these expressions and adding them, we obtain

$$A'^2 + B'^2 = A^2 + B^2$$

Thus we see that under the operation of translation the *pair* of functions $\cos x$
and $\sin x$ constitutes the eigenfunctions, for when placed into the operation of
translation by the amount h, they again emerge.

The Euler identities

$$\cos x + i \sin x = e^{ix}$$
$$\cos x - i \sin x = e^{-ix}$$

(where $i = \sqrt{-1}$) lead to the corresponding formulas

$$\cos x = \frac{1}{2}(e^{ix} + e^{-ix})$$

$$\sin x = \frac{1}{2i}(e^{ix} + e^{-ix})$$

In this notation the two addition formulas from trigonometry are contained in the single, much simpler formula

$$e^{ix}e^{iy} = e^{i(x+y)}$$

This fact can be seen by using the Euler identities and equating the real and imaginary terms on both sides. Thus the complex exponential is the eigenfunction of translation. It is far more convenient, therefore, to use complex exponentials than the real sines and cosines.

We have chosen the mathematical convention of $i = \sqrt{-1}$ rather than the engineering convention of j. The choice is arbitrary, but since the book is designed mainly for nonengineers, i is a reasonable choice.

It is natural to ask if the sine and cosine are unique in having this property of invariance under translation. The invariant property that we want is that both functions under a translation of a fixed amount, say h, can be written as a linear combination of sine and cosine or, more generally, that any linear combination of a sine and cosine of a given frequency can be written as a linear combination when an arbitrary translation of size h of the coordinate axis is made. We are further assuming that the functions are odd and even respectively, and that the trigonometric functions, sine and cosine, are reasonably smooth. Under these assumptions it can be shown that the sines and cosines are unique (except that the corresponding hyperbolic functions are also possible). Notice that in the complex exponential notation the eigenfunction property is much more simply expressed than in the trigonometric notation,

$$y(t + h) = e^{i\omega(t+h)} = e^{i\omega h}e^{i\omega t} = \lambda(\omega)y(t)$$

where $\lambda(\omega)$ is the eigenvalue

$$\lambda(\omega) = e^{i\omega h}$$

and is independent of the variable t.

2.5   LINEAR SYSTEMS

The second eigenfunction property that we wish to show is that the complex exponential functions $e^{i\omega t}$ and $e^{-i\omega t}$ are eigenfunctions for linear, time-invariant systems. In abstract notation this means

$$L\{e^{i\omega t}\} = \lambda(\omega)e^{i\omega t}$$

where $L\{\ \}$ is an arbitrary linear operator. A linear operator has the property that

$$L\{ag_1(t) + bg_2(t)\} = aL\{g_1(t)\} + bL\{g_2(t)\}$$

Clearly, for nonrecursive filters of the form

$$y_n = \sum_{k=-k}^{K} c_k x_{n-k}$$

the substitution

$$x(t) = e^{i\omega t}$$

produces, when we factor out the exponential term depending on n, the output

$$y(n) = e^{i\omega n} \sum_{k=-K}^{K} c_k e^{-i\omega k} = \lambda(\omega)e^{i\omega n}$$

where

$$\sum_{k=-K}^{K} c_k e^{-i\omega k} = \lambda(\omega)$$

Thus the function $e^{i\omega t}$, which we put into the right-hand side of the equation, can be factored out of the expression and appears multiplied by its eigenvalue $\lambda(\omega)$. The eigenvalue $\lambda(\omega)$ is, of course, a constant as far as t or, equivalently, n is concerned.

It is easy to see that a similar situation applies to recursive filters. We have only to substitute complex exponentials with the same frequencies, but possibly different amplitudes, for *both* x_k and y_k and note that the result is an expression independent of n.

It is worth noting that the exponential function is also the eigenfunction that is appropriate for the calculus operations of differentiation

$$\frac{d}{dt} e^{i\omega t} = i\omega e^{i\omega t}$$

and integration

$$\int e^{i\omega t}\, dt = \frac{e^{i\omega t}}{i\omega}$$

The exponential is also the eigenfunction for differencing, since

$$\Delta e^{i\omega t} = e^{i\omega(t+1)} - e^{i\omega t} = e^{i\omega t}[e^{i\omega} - 1]$$

Thus we see, contrary to the impression gained from the usual calculus course, that the powers of x are not the eigenfunctions of calculus. Instead the exponentials, real or complex, are the natural, the characteristic, the eigenfunctions of the calculus.

Exercises

2.5-1   Find the eigenvalue corresponding to the kth derivative.

2.5-2   Find the eigenvalue corresponding to the kth difference operator Δ^k.

2.5-3   Carry out the details for the recursive filter equation.

2.6   THE EIGENFUNCTIONS OF EQUALLY SPACED SAMPLING

The purpose of this section is to show that the eigenfunctions of the process of equally spaced sampling of a function are the common sines and cosines of trigonometry. The sense in which we mean that they are eigenfunctions is that when we (a) take a sinusoid of some frequency (think of it as a high frequency), then (b) do the process of sampling the given frequency function at equally spaced points, and (c) finally ask "What equivalent sinusoid of low frequency do we have?" we find that it is equivalent to a *single* sinusoid function. Stated simply, aliasing takes any particular frequency and, in the sense of having the same values at the sample points, transforms it into a single low-frequency function.

Let us contrast this result with what happens when the classical polynomial method of approximation is used. In polynomial approximation using the sample points x_i ($i = 1, \ldots, N$), we are led directly to consider the *sample polynomial* defined by

$$\pi(x) = [x - x_1][x - x_2]\cdots[x - x_N]$$

This is the function that vanishes at all the sample points x_i and thus is the function that we cannot "see." Now given any power of x, say x^m, we divide this power by $\pi(x)$ in order to get a quotient $Q(x)$ and a remainder $R(x)$,

$$x^m = \pi(x)Q(x) + R(x)$$

where $R(x)$ is of degree less than N. A simple generalization of the standard remainder theorem shows that *at the sample points* x_i the two functions x^m

and $R(x)$ have exactly the same values. Thus the original single power of x is aliased into a polynomial $R(x)$, which is, of course, a linear combination of $1, x, x^2, \ldots, x^{(N-1)}$ and not a single power. In this sense, the powers of x are not eigenfunctions for sampling at any spacing. Aliasing for polynomials is a messy business.

Let us restate this result. If we regard the process as (a) starting with a basis function (a power of x, say x^m), (b) sampling at $n + 1$ points, and finally, (c) constructing from the samples a new function of minimal degree in x, then we see that, in general, a single power of x does not go into a power of x. On the other hand, for sinusoids, the process of equally spaced sampling followed by the reconstruction of a function of minimal frequency does result in a single sinusoid. Consequently, in this sense, the sinusoids are the eigenfunctions of equally spaced sampling and the process reveals once more the central role that aliasing plays in the equally spaced sampling process.

2.7   SUMMARY

In this chapter we have given three reasons why the trigonometric functions, sine and cosine, are the eigenfunctions to be used in many filter problems, discrete or continuous. In particular, they are the eigenfunctions for

Invariance under translation by an arbitrary amount

Linear systems

Equally spaced sample systems

In the form $e^{i\omega t}$ and $e^{-i\omega t}$, the trigonometric functions are more easily handled in many problems; but it should be remembered that we are modeling a real world and that it is, in the final analysis, a real signal that we record and transform—in spite of the great convenience of the complex notation. In the complex notation we also see why there are both positive and negative frequencies. Corresponding to the two real functions sine and cosine at a single positive frequency, we have the two functions $e^{i\omega t}$ and $e^{-i\omega t}$.

3

Some Classical
Applications

3.1 INTRODUCTION

This chapter uses the frequency approach to examine a number of classical applications of digital filters (although they are usually not presented as digital filters). Looking first at familiar situations makes the frequency approach both easier to understand and psychologically more acceptable. The original derivation of each case was based on the classical polynomial approach to numerical methods which uses polynomials as the basis for approximation. By making this comparison, the frequency approach will be shown to shed new light on well-known situations.

Although long, the chapter simply presents a sequence of applications using the same set of ideas. In a sense, the reader is being overwhelmed with particular cases. Hopefully, a few of them will be sufficiently familiar so that the new approach does produce significant new insights. Thus the size of the chapter is due more to the variety of assumed backgrounds of the readers than to the importance of the individual examples.

Having mastered this new method of analyzing and evaluating linear formulas by using complex exponentials, it is natural to turn to the corresponding design problem. To do this we will need the formal mathematics of Fourier series, which is discussed in Chapters 4 and 5.

3.2   LEAST-SQUARES FITTING OF POLYNOMIALS

Suppose that we have M data points (t_m, x_m), $m = 1, 2, \ldots, M$, and wish to fit (approximate) this data, in some sense, by a polynomial $x = x(t)$ of degree N, where $M > N + 1$ (meaning that there are more data points than there are parameters in the polynomial). See Fig. 3.2-1. In general, we cannot hope to find coefficients so that the polynomial exactly fits all the data (even neglecting computer roundoff). The errors of the polynomial fit are called the *residuals*, and, as a rule, they will not all be zero. *The principle of least squares* states that of all polynomials of degree N we should select the one for which the sum of the squares of the *residuals* is the least. This principle is, of course, an assumption and should not be taken as infallible truth.

As an example, consider a set of five equally spaced data points. For ease of finding the formula, we fix the coordinate system and pick the points at $t_m = m$ for $m = -2, -1, 0, 1, 2$, with the corresponding values of x_m unspecified. If we fit this data by a straight line $x = A + Bt$, in a least-squares sense, then we minimize

$$F(A, B) = \sum_{m=-2}^{2} [x_m - (A + Bm)]^2$$

The variables of the problem are the coefficients A and B of the straight line. To find a minimum, we naturally differentiate with respect to A and B and then set the resulting expressions equal to zero. This process gives the following two equations (since $\sum t_m = 0$ and $\sum t_m^2 = 10$)

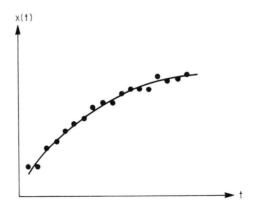

FIGURE 3.2-1   FITTING DATA LEAST SQUARES

$$\begin{cases} 5A + 0B = \sum_{m=-2}^{2} x_m \\ 0A + 10B = \sum_{m=-2}^{2} m x_m \end{cases}$$

These equations are called *the normal equations*, presumably because they normally arise in least-squares polynomial fitting. Before solving them for both A and B, what do we want? In the common case of smoothing, we use the midpoint of the line as the smoothed value $\bar{x}$ in place of the original data point x_0. In this case, we need find only A, which is, from the first normal equation, one-fifth of the sum of the five data values. Here the smoothed value is the mean.

We fitted the straight line to the five points, but typically, when smoothing data, we fit a straight line to each (overlapping) set of five adjacent points and take the corresponding smoothed values $\bar{x}$ as the values at the midpoint of the corresponding five points. When going to the moving coordinate system to do this smoothing, we find that we are smoothing by 5s and that we have the formula

$$\bar{x}_n = \frac{1}{5} \sum_{m=-2}^{2} x_{n-m}$$

where x_n is the smoothed value at $t = t_n$. The five coefficients c_m of the filter are each $\frac{1}{5}$. The noise amplification factor (Section 1.7) is also $\frac{1}{5}$. Note that we are not able, with this formula, to smooth the two values at the start of the run of data we are given and the two values at the end of the date.

How does this formula look from the frequency point of view? Suppose that the input function is the complex sinusoid $e^{i\omega t}$. Since the formula is linear in the data, we know (from Section 2.5) that the same function will emerge *except* that it is multiplied by its eigenvalue, which we will label from now on as

$$H(\omega)$$

The eigenvalue depends on ω but not on t (in this case, n). Putting the complex exponential $e^{i\omega t}$ into the smoothing by 5s formula, we obtain

$$H(\omega) = \frac{1}{5}[e^{-2i\omega} + e^{-i\omega} + 1 + e^{i\omega} + e^{2i\omega}]$$

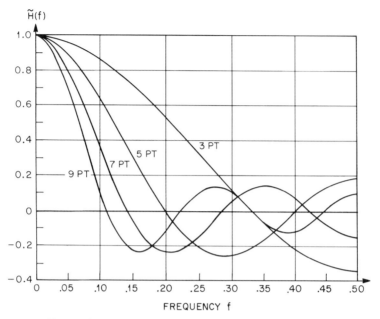

FIGURE 3.2-2   SMOOTHING BY LEAST SQUARES STRAIGHT LINES

Note that the negative and positive frequencies have the same coefficient, and therefore they can be replaced by the corresponding cosine functions

$$H(\omega) = \frac{1}{5}[1 + 2\cos\omega + 2\cos 2\omega].$$

This function $H(\omega)$, which is the eigenvalue of the formula, is called *the transfer function*, since it is what enables us to take the input frequency and *transfer* to the output by simply multiplying by $H(\omega)$. This curve, along with several to be discussed in the following paragraphs, is shown in Fig. 3.2-2 and is labeled 5-point. Here we have used frequency f as the independent variable, where $f = \omega/2\pi$.

An alternate representation of the transfer function is based on the fact that in the complex exponential form it is a geometric progression with $r = e^{i\omega}$,

$$H(\omega) = \frac{e^{5i\omega/2} - e^{-5i\omega/2}}{5[e^{i\omega/2} - e^{-i\omega/2}]} = \frac{\sin(5\omega/2)}{5\sin(\omega/2)}$$

The transfer function, $H(\omega)$, is clearly a periodic function of ω. However, because of the original sampling at unit spacing, it has meaning only in an interval of length 2π, which we conventionally take as running from $-\pi$ to

π. To extend the transfer function beyond this range is to confuse frequencies that are already aliased by the sampling into the fundamental interval with those that before the sampling lay outside. Thus we shall always draw at most one period of the transfer function. This cutoff frequency, where the aliasing due to the sampling process occurs, is called the *folding* or *Nyquist frequency*.

For convenience, we always plot our figures in terms of the rotational frequency f rather than the angular frequency ω. The quantities f and ω are related by the formulas

$$2\pi f = \omega \quad \text{or} \quad f = \frac{\omega}{2\pi}$$

Thus $H(\omega) = H(2\pi f)$. Unfortunately, this latter function is often written $H(f)$, which is a source of confusion. We will therefore write

$$H(2\pi f) = \tilde{H}(f)$$

It is the same confusion that occurs, say, with the sine function when radians and degrees are both used in the same discussion. Figure 3.2-2 shows graphically what happens at any individual frequency f: an input amplitude is simply multiplied by the eigenvalue $\tilde{H}(f)$ plotted in the figure. The lowest frequency $f = 0$ (in electrical applications this corresponds to direct current and is often abbreviated dc) is transmitted through the smoothing filter with the correct amplitude. All other frequencies suffer some attenuation (decrease). The smoothing by $5s$ has two frequencies, $f = 2/10$ and $4/10$, such that the amplitude of the output is zero regardless of the amplitude of the corresponding input frequency.

If we fit a straight line to $2m + 1$ points and proceed as we did with 5 points, then we are led to the smoothing by the $(2m + 1)$-point formula. The corresponding transfer function $H(\omega)$ gives the eigenvalue for frequency ω

$$H(\omega) = \frac{1}{(2m + 1)}[1 + 2 \cos \omega + \cdots + 2 \cos m\omega]$$

The alternate representation is

$$H(\omega) = \frac{\sin (m + 1/2)\omega}{(2m + 1) \sin (\omega/2)}$$

From this formula we see that the more terms used in the smoothing formula, the more rapid are the wiggles of the transfer function $H(\omega)$ and the more the envelope of the wiggles is squeezed toward the frequency axis. For the graphs of the first few cases, $2m + 1 = 3, 5, 7, 9$, see Fig. 3.2-2. Notice that the sin $(\omega/2)$ in the denominator goes from 0 to 1 as ω goes from 0 to π (f goes from 0 to $\frac{1}{2}$).

These smoothing formulas are clearly the same as the running *average* in statistics, and thus the transfer function $H(\omega)$ shows what happens when data of frequency ω is averaged. Notice that these transfer functions are periodic and symmetric about both $\omega = 0$ and $\omega = \pi$ ($f = \frac{1}{2}$). The periodicity about $f = \frac{1}{2}$ reflects the aliasing that we earlier found must occur when sampling at equally spaced points. The periodicity about $f = 0$ arises from the symmetry of the coefficients of the formula.

Notice that since we are using eigenfunctions, the operation of smoothing does not cause the various terms to interact. So if the input function $x(t)$ is a sum of M complex sinusoids

$$x(t) = \sum_{m=1}^{M} C_m e^{i\omega_m t}$$

then the output is

$$\sum_{m=1}^{M} C_m H(\omega_m) e^{i\omega_m t}$$

Exercises

3.2-1   Construct (real-valued) data corresponding to $f = 4/10$ and smooth it by 5s.

3.2-2   Derive the formula for smoothing by $(2m + 1)$-point moving average.

3.2-3   Using only trigonometry, show that the two formulas in the text for smoothing by $(2m + 1)$ points are the same.

3.2-4   Give explicit formulas for the zero crossings of the transfer functions in Fig. 3.2-2 and for the values at $f = \frac{1}{2}$.

3.3   LEAST-SQUARES QUADRATICS AND QUARTICS

Instead of smoothing by fitting a straight line, suppose that we fit a quadratic (or a cubic, which is equivalent). For the quadratic

$$x(t) = A + Bt + Ct^2$$

we minimize the sum of squares of the residuals,

$$F(A, B, C) = \sum [x_m - (A + Bm + Cm^2)]^2$$

By differentiating with respect to the variables of the fit—namely, A, B, and C—we obtain the normal equations

$$\begin{cases} A\{1\} + B\{m\} + C\{m^2\} = \{x\} \\ A\{m\} + B\{m^2\} + C\{m^3\} = \{mx\} \\ A\{m^2\} + B\{m^3\} + C\{m^4\} = \{m^2x\} \end{cases}$$

where $\{\cdot\}$ means "sum over the values of the argument enclosed in the braces." By selecting the particular set of equally spaced data points $-m$, $-(m-1), \ldots, m$, the sums over the odd powers are all zero. Solving for A, which is the only value needed, we have

$$A = \frac{[\{m^4\}\{x\} - \{m^2\}\{m^2x\}]}{[\{1\}\{m^4\} - \{m^2\}^2]}$$

Written out for the case of 5 points, this is

$$\bar{x}_0 = A = \frac{\left[34 \sum_{m=-2}^{m=2} x_m - 10 \sum_{m=-2}^{m=2} m^2 x_m \right]}{(5)(34) - (10)(10)}$$

$$= \frac{-3x_{-2} + 12x_{-1} + 17x_0 + 12x_1 - 3x_2}{35}$$

Here the smoothed value is a weighted average of the five input values. For the general point we have

$$\bar{x}_n = \frac{-3x_{n-2} + 12x_{n-1} + 17x_n + 12x_{n+1} - 3x_{n+2}}{35}$$

To analyze this formula, we substitute

$$x_m = e^{i\omega m}$$

The formula becomes

$$\frac{17 + 24 \cos \omega - 6 \cos 2\omega}{35}$$

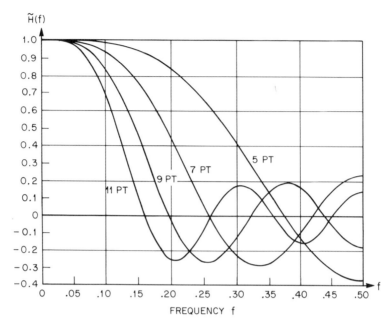

FIGURE 3.3-1 TRANSFER FUNCTION FOR SMOOTHING BY LEAST-SQUARES QUADRATICS

and is, of course, our $H(\omega)$. The transfer functions $H(\omega)$ of these curves are shown in Fig. 3.3-1 for 5, 7, 9, and 11 points. The coefficients c_k are, respectively,

$$\frac{1}{35}[-3, 12, 17, 12, -3]$$

$$\frac{1}{21}[-2, 3, 6, 7, 6, 3, -2]$$

$$\frac{1}{231}[-21, 14, 39, 54, 59, 54, 39, 14, -21]$$

$$\frac{1}{429}[-36, 9, 44, 69, 84, 89, 84, 69, 44, 9, -36]$$

Remember that in the transfer function the noncentral coefficients must be doubled in order to get the corresponding coefficients of the cosines. In Fig. 3.3-1 we again plot the independent variable in terms of the rotational frequency f.

These curves of $\tilde{H}(f)$ describe the transfer function of least-squares smoothing by quadratics and resemble those for smoothing by straight lines *except* for the higher degree of tangency at $\omega = 0$. In both cases, using more terms in the smoothing formula causes the curves to come down more rapidly and to be slightly lower in the subsequent wiggles.

Continuing, we go to the next case and smooth by fitting least-squares quartics (quintics give the same results). We therefore assume a quartic of the form

$$x(t) = A + Bt + Ct^2 + Dt^3 + Et^4$$

Then using the data points $t = -m, -(m-1), \ldots, 0, 1, \ldots, m$ (which temporarily fixes the coordinate system and makes the notation much simpler), we set up the differences (the residuals) between the data and the calculated values, square them, and sum over all the data. The next step is to seek the minimum of this function of the variables A, B, C, D, E. As usual, the function to be minimized is differentiated in turn with respect to each of the variables, and the corresponding derivatives are set equal to zero. The result is the normal equations. Only the first, third, and fifth of these equations are necessary, since all we need is the value A. For $m = 3, 4, 5, 6$ or, what is the same thing, for 7, 9, 11, 13 points, we get the value of A, *which will be linear in the original data* x_m. The smoothing formulas obtained have the following coefficients c_k:

$$\frac{1}{231}[5, -30, 75, 131, 75, -30, 5]$$

$$\frac{1}{429}[15, -55, 30, 135, 179, 135, 30, -55, 15]$$

$$\frac{1}{429}[18, -45, -10, 60, 120, 143, 120, 60, -10, -45, 18]$$

$$\frac{1}{2431}[110, -198, -135, 110, 390, 600, 677, 600, 390, 110, -135, -198, 110]$$

As usual, we assume the input function $x(t) = e^{i\omega t}$ and obtain the transfer functions $H(\omega)$. In these cases, they are shown in Fig. 3.3-2, where we again use angular frequency f in plotting.

Once more the effect of the higher-degree polynomial is a higher degree of tangency at $\omega = 0$; in addition, the use of extra terms in the smoothing formula makes the curve come down sooner.

This higher degree of tangency at $\omega = 0$ is a general result.

THEOREM: *The more powers of t that we fit in the time domain, the higher the tangency at* $\omega = 0$ *in the frequency domain.*

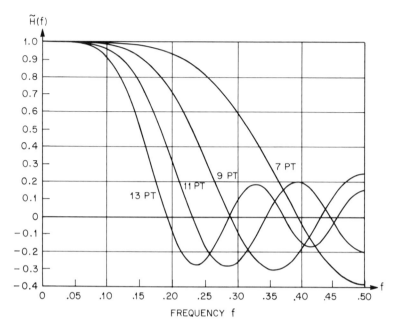

FIGURE 3.3-2   TRANSFER FUNCTION FOR SMOOTHING BY LEAST-SQUARES
QUARTICS

To prove this theorem, we assume a smoothing formula of the form

$$\bar{x}_n = \sum_{k=-K}^{k=K} c_k x_{n-k}$$

Assume also that the coefficients c_k have been determined to make the formula
true for $x(t) = 1, t, t^2, \ldots, t^{(m-1)}$ but not true for t^m. This means that the
equations (using $n = 0$)

$$\sum_{k=-K}^{k=K} k^p c_k = \begin{cases} 1, & p = 0 \\ 0, & p = i, \ldots, m-1 \\ \neq 0, & p = m \end{cases}$$

Now consider the transfer function in the form

$$H(\omega) = \sum_{k=-K}^{k=K} c_k e^{-ik\omega} = \sum_{k=-K}^{k=K} c_{-k} e^{ik\omega}$$

Since the equation is true for $x(t) = 1$, it follows that the preceding equation is true for $\omega = 0$, and therefore $H(0) = 1$. By differentiating with respect to ω and then setting $\omega = 0$, we get the equation

$$\sum_{k=-K}^{k=K} k c_k = 0$$

From this it follows that $H'(0) = 0$. Repeated differentiation and setting $\omega = 0$ will show that the successive derivatives are zero, up through the $(n - 1)$st derivative. But for the nth derivative there will not be cancellation of the terms, and so that corresponding derivative of $H(\omega)$ will not be zero. Thus we have proved the theorem.

Exercises

3.3-1   Derive in detail the transfer function for the 5-point least-squares quartic. Give the answer in terms of consines.

3.3-2   Derive the transfer function for the 7-point quartic. Give the answer in terms of cosines.

3.4   INTEGRATION: RECURSIVE FILTERS

Another well-known operation that uses a linear combination of the data is *numerical integration*; in particular, the reader is probably familiar with the trapezoid rule, the midpoint rule, and Simpson's formula. Let us examine these formulas from the frequency point of view.

The trapezoid rule is (using $y_0 = 0$)

$$y_{n+1} = y_n + \frac{1}{2}[x_{n+1} + x_n]$$

where the x_n are the integrand values and the y_n are the integral values (area values). Since y occurs on both sides of this equation, this filter is clearly recursive (Section 1.1). On the other hand, the earlier smoothing formulas considered were nonrecursive filters. In this case, we assume that the input $x(t)$ is $e^{i\omega t}$ and, since the equation is linear, that the corresponding output $y(t)$ is of the form $A(\omega)e^{i\omega t}$ (Section 2.5). Solving for $A(\omega)$, we obtain

$$A(\omega) = \frac{1}{2}\frac{[e^{i\omega} + 1]}{[e^{i\omega} - 1]}$$

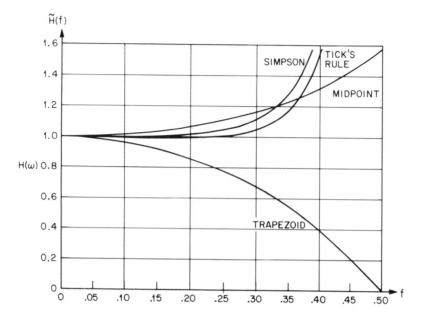

FIGURE 3.4-1 FREQUENCY RESPONSE OF INTEGRATION FORMULAS

Divide numerator and denominator by $e^{i\omega/2}$ and write the result in terms of the conventional trigonometric functions

$$A(\omega) = \frac{\cos \omega/2}{2i \sin \omega/2}$$

The true answer for integration of the function $e^{i\omega t}$ is, of course, $[1/i\omega]e^{i\omega t}$. We now take the ratio of the calculated to true

$$\frac{\text{calculated}}{\text{true}} = \cos \frac{\omega}{2} \left[\frac{(\omega/2)}{\sin (\omega/2)} \right]$$

At $\omega = 0$, this ratio is clearly one and falls off to zero at the Nyquist frequency, $\omega = \pi$ ($f = \frac{1}{2}$), which is the natural boundary, for it is there that aliasing begins. See Fig. 3.4-1.

Why have we taken the ratio of the computed to true answers instead of simply evaluating the transfer function as in the past? The answer is simple; for both smoothing and filtering, we took the ratio of the output to the input, which was the natural comparison to see how well we did. But in operations like differentiation and integration we compare the result obtained with the result wanted, and so the ratio of the calculated to true values is used for evaluating such formulas. Clearly, the right-hand side is *a transfer function* that we can use to go (transfer) from the true value to the calculated value.

Expanding both numerator and denominator in power series and then dividing them out, we have

$$\frac{\text{calculated}}{\text{true}} = 1 - \frac{\omega^2}{12} + \frac{\omega^4}{720} + \cdots$$

which gives the shape of the ratio near $\omega = 0$.

This same process on Simpson's formula for integration (using $y_0 = 0$)

$$y_{n+1} = y_{n-1} + \frac{1}{3}[x_{n+1} + 4x_n + x_{n-1}]$$

gives

$$\frac{\text{calculated}}{\text{true}} = \frac{[2 + \cos \omega]}{[3(\sin \omega)/\omega]} = 1 + \frac{\omega^4}{180} + \cdots$$

This curve begins at $\omega = 0$ with the value 1, has (according to the earlier theorem of Section 3.3) a tangency through the third derivative, and rises until at $\omega = \pi$ the denominator becomes zero. (See Fig. 3.4-1.)

Since this behavior at $\omega = \pi$ is surprising, how can we understand it? Clearly, Simpson's formula amplifies the upper part of the Nyquist interval (the higher frequencies), whereas the trapezoid rule damps them out. At the Nyquist frequency the function being integrated could have the values $+1$, $-1, +1, -1, +1, -1, \ldots$, and in Simpson's formula these values are going to be multiplied by the coefficients 1, 4, 2, 4, 2, 4, 2, 4, $\ldots$, all divided by 3, of course. Looking at the products, we see that each pair of values, the second and third (-1×4 and $+1 \times 2$), the fourth and fifth, and so on, combines to give a negative number; thus for this high-frequency function the integrated sum will grow in size in a linear fashion. This effect is known in numerical analysis but is seldom mentioned in textbooks.

The corresponding effect for the trapezoid rule will not show this effect. For frequencies near the Nyquist frequency, we see that long stretches of values will not tend to combine as they do for Simpson's integration formula.

Next, we look at the midpoint integration formula (using $y_0 = 0$)

$$y_{n+1} = y_n + x_{n+1/2}$$

The result of substituting $e^{i\omega t}$ for $x(t)$, $A(\omega)e^{i\omega t}$ for $y(t)$, and computing the ratio of the calculated to true values, is

$$\frac{\omega/2}{\sin(\omega/2)}$$

This expression begins at one and slowly rises as ω increases. At the Nyquist limit, we have $\pi/2$ as the value of the function.

The fourth curve of Fig. 3.4-1 is the transfer function of a formula due to Leo Tick. The formula was designed to have a transfer function that is as close to unity as we can make it throughout the lower half of the Nyquist interval while still involving only three consecutive terms. The formula is (using $y_0 = 0$)

$$y_{n+1} = y_{n-1} + h(0.3584x_{n+1} + 1.2832x_n + 0.3584x_{n-1})$$

We will derive this formula in Section 12.10.

In summary, we have shown how three classical numerical integration formulas look when examined from the frequency point of view. The use of the natural eigenfunctions of the problem $e^{i\omega t}$ has, as expected, shed a new and different light on each of them. The implications of what the formulas do to noisy data was intuitively understood in hand-calculating days but has not, generally speaking, appeared in modern textbooks. It is clear that in the presence of noise, which usually includes a good deal of high frequency, Simpson's formula is more dangerous to use than the trapezoid or midpoint formulas. But when there is relatively little high frequency in the function being integrated, then the flatness of Simpson's formula for low frequencies shows why it is superior. The choice to make (once shown) depends in an obvious way on the frequency characteristics of the function being integrated.

A word of caution, however. Actually, it is not possible to judge a method of computation without considering what is to be done with the results. If, for instance, the integrated function is to be analyzed for frequencies, then it is particularly important to examine the effects of the method of integration on various frequencies and to make allowances in the interpretation for these effects. As noted, Tick's integration rule of Fig. 3.4-1 was designed to be as accurate as it could be (given the basic form) for the lower half of the Nyquist interval, $0 \leq f \leq \frac{1}{4}$. The figure shows (just barely) that the error at the upper limit of the band in which it was to be accurate is the same as the maximum error at any earlier point except that the sign is different. Thus the maximum error for any frequency in this interval is minimized. This criterion will be studied more closely in Chapter 12. The price of this accuracy over a large range of frequencies is, among other factors, less accuracy near $\omega = 0$.

Exercises

3.4-1   Apply the methods of this section to the three-eighths rule of integration

$$y_{n+2} = y_{n-1} + \frac{1}{8}[x_{n+2} + 3x_{n+1} + 3x_n + x_{n-1}]$$

3.4-2   Compute and plot the transfer functions for the Newton-Cotes integration formulas for $N = 2, 3, \ldots, 10$ [*H*, p. 342].

3.4-3   Compute the sum of the squares of the coefficients of the x_n terms (the noise amplification of Section 1.7) for the four integration formulas of this section.

3.5   DIFFERENCES AND DERIVATIVES

The difference operator

$$\Delta y_n = y_{n+1} - y_n$$

provides another class of examples showing the advantages of the frequency approach to a polynomial-derived formula. The repeated use of this operator Δ leads to

$$\Delta^k[y_n] = \Delta[\Delta^{(k-1)}][y_n]$$

Its importance arises from the theorem that the operator Δ^{n+1} annihilates a polynomial $P_n(x)$ of degree n in x; that is,

$$\Delta^{(n+1)}P_n(x) = 0$$

Furthermore, the difference operator can be shown to "amplify" small errors [*H*, Chapters 10 and 35]. Thus the difference table of a function that is identically zero but that has a single error (because of linearity, we can take it to be 1) will have the binomial coefficients with alternating signs in the successive columns of differences.

Using the frequency approach, we now examine the effect that the opera-

TABLE 3.5-1 Difference table

n	$f(n)$	$\Delta f(n)$	$\Delta^2 f(n)$	$\Delta^3 f(n)$
-3	0			
		0		
-2	0		0	
		0		1
-1	0		1	
		1		-3
0	1		-2	
		-1		3
1	0		1	
		0		-1
2	0		0	
		0		0
3	0		0	
		0		
4	0			

tor Δ has on an arbitrary frequency $e^{i\omega t}$; we have

$$\Delta e^{i\omega t} = e^{i\omega(t+1)} - e^{i\omega t}$$
$$= [e^{i\omega} - 1]e^{i\omega t}$$
$$= e^{i\omega/2}[e^{i\omega/2} - e^{-i\omega/2}]e^{i\omega t}$$
$$= ie^{i\omega/2}\left[2\sin\frac{\omega}{2}\right]e^{i\omega t}$$

It follows that k repeated application of the Δ operator will give

$$\Delta^k[e^{i\omega t}] = [i^k]e^{ik\omega/2}\left[2\sin\frac{\omega}{2}\right]^k e^{i\omega t}$$

The first two factors are of size 1, and so the *amplification* at the frequency ω is contained in the factor

$$\left[2\sin\frac{\omega}{2}\right]^k$$

where $-\pi \leq \omega \leq \pi$ is the usual Nyquist interval. We see immediately that for the smallest one-third of the frequencies, $0 \leq \omega < \pi/3$, the difference operator Δ decreases the amplitude of any frequency, whereas in the upper two-thirds of the frequencies, $\pi/3 < \omega \leq \pi$, there is amplification. See Fig. 3.5-1. This situation explains the typical use of the difference table to locate

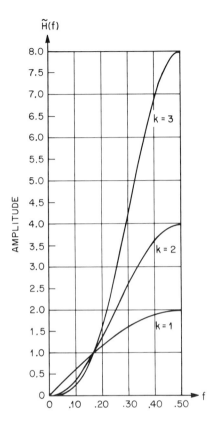

FIGURE 3.5-1   FREQUENCY RESPONSE OF DIFFERENCE OPERATOR Δ^k

(high-frequency) noise. Noise in this case means the upper two-thirds of the Nyquist interval.

Differences are also used to approximate derivatives. For instance, the central difference formula

$$x_{n+1} - x_{n-1} = 2hx'_n$$

provides an approximation to the derivative. Using our frequency approach, we set

$$x(t) = e^{i\omega t}$$

The true derivative is

$$x'(t) = i\omega e^{i\omega t}$$

From the formula (with $h = 1$) the calculated value is

$$e^{i\omega} - e^{-i\omega} = 2i \sin \omega$$

The ratio of the calculated to the true answer is

$$\frac{2i \sin (\omega)}{2i\omega} = \frac{\sin \omega}{\omega}$$

Thus when $\omega = 0$ (dc), the value of the ratio is 1, as it should be, but for all other ω the ratio is less than 1. Thus the formula underestimates the value of the derivative for all other frequencies in the Nyquist interval.

For the second derivative, we use the estimate

$$x''(t) = x(t + 1) - 2x(t) + x(t - 1)$$

and obtain the ratio of the calculated to true answers as

$$\frac{\text{calculated}}{\text{true}} = \frac{[e^{i\omega} - 2 + e^{-i\omega}]}{(i\omega)^2}$$

$$= \frac{2(1 - \cos \omega)}{\omega^2}$$

$$= \left[\frac{\sin (\omega/2)}{\omega/2} \right]^2$$

which is the square of the preceding curve but contracted by a factor of 2 in the independent variable.

The beginner should take some particular frequency and see how these formulas agree with practice by operating on definite numbers and, in particular, notice how the zero estimates arise.

Exercises

3.5-1   Apply the theory of the preceding section to the following third derivative approximation $[-x(n + 2) + 2x(n + 1) - 2x(n - 1) + x(n - 2)]/2$

3.5-2   Discuss the tangency of the transfer function of the kth derivative at $\omega = 0$.

3.6   MORE ON SMOOTHING: DECIBELS

Sections 3.2 and 3.3 show that many smoothing formulas using least squares keep the value at zero frequency (dc) correct but, in general, decrease the amount of any higher frequency that may be in the function being smoothed. The curves in the figures are the *transfer functions of the linear process of smoothing*; that is, they are, for each ω, the corresponding eigenvalue of the process in the range up to the Nyquist frequency (which is where aliasing begins).

This situation suggests examining other classical smoothing formulas in order to see if they have a similar property. Perhaps the two best-known smoothing formulas that have come down to us from the past are Spencer's 15-and 21-point smoothing formulas [KS3, p.372]

$$x_n = \frac{1}{320}[-3x_{n-7} - 6x_{n-6} - 5x_{n-5} + 3x_{n-4} + 21x_{n-3}$$
$$+ 46x_{n-2} + 67x_{n-1} + 74x_n + 67x_{n+1} 46x_{n+2} + 21x_{n+3}$$
$$+ 3x_{n+4} - 5x_{n+5} - 6x_{n+6} - 3x_{n+7}]$$

and

$$x_n = \frac{1}{350}[-x_{n-10} - 3x_{n-9} - 5x_{n-8} - 5x_{n-7} - 2x_{n-6}$$
$$+ 6x_{n-5} + 18x_{n-4} + 33x_{n-3} + 47x_{n-2} + 57x_{n-1}$$
$$+ 60x_n + 57x_{n+1} + \cdots]$$

The transfer functions for these two formulas are given in Fig. 3.6-1. If we accept the hypothesis that they were also designed to remove "noise," then we are forced to the conclusion that "noise" was classically identified with

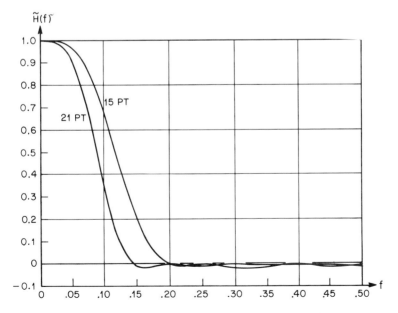

FIGURE 3.6-1   TRANSFER FUNCTION OF SPENCER'S SMOOTHING FORMULAS

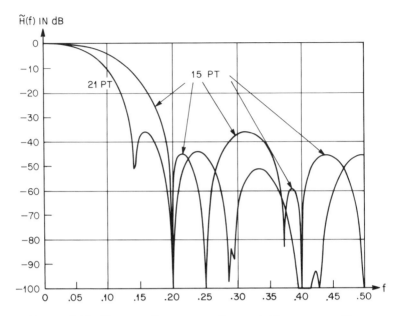

FIGURE 3.6-2 TRANSFER FUNCTION OF SPENCER'S FORMULAS IN dB

high frequencies and "message" or "information" with low frequencies. The two formulas differ in how much they let pass through the filtering (smoothing) process. The longer formula has the shorter pass band of frequencies and is about what would be expected from general experience in science.

These formulas were designed not only to pass low frequencies and stop high frequencies but also to be easy to compute by hand. Consequently, they are not necessarily optimal for present computers. Indeed, for the number of terms and the given "cutoff frequency," better formulas can easily be designed (see Chapter 6 and 9).

As shown in Fig. 3.6-1, the curves are not too informative, since they are so small at the higher frequencies that we cannot see how good they are. It is therefore better to plot the logs of the numbers $H(\omega)$. For this purpose, it is customary to use decibels (tenths of a bel), abbreviated dB and defined as

$$20 \log (\text{ratio}) = \text{dB units}$$

where, of course, we pick the reference value (the denominator of the ratio) as the original amount of the frequency present in the signal. Figure 3.6-2 shows this version of the Spencer's smoothing formulas. The past inference that noise was high frequency and signal low frequency is strengthened as we examine this new plot of the corresponding transfer functions.

It is now clear that smoothing formulas generally remove some frequencies and let other pass. In 1927 Slutsky and Yule both reported on this effect [KS3, p.378]—namely, that smoothing could greatly effect what was later found by further analysis, especially when examining data with large amounts of noise. As a result of this filtering effect, sometimes the periods that are found in smoothed data are more the effect of the smoothing than of the original data. (See Section 13.4)

3.7 MISSING DATA AND INTERPOLATION

In a long record of data there are often one or more (isolated) missing values. This situation occurs for various reasons: the measurements may never have been made; they may have been misrecorded and thus later removed; or the formula used to compute the successive values of the function may have involved an indeterminate form, such as $(\sin x)/x$ at $x = 0$, and the computer refused to divide by zero.

Perhaps the most common way of supplying an isolated missing value is to use an interpolation formula based on the assumption that the data locally is a polynomial of some odd degree. This is equivalent to the assumption that the next higher-order difference is zero. For instance, assume that the fourth difference is zero. We therefore have

$$\Delta^4 x_{n-2} \equiv x_{n-2} - 4x_{n-1} + 6x_n - 4x_{n+1} + x_{n+2} = 0$$

Solving this equation for x_n gives the very convenient, stable formula for the missing value

$$x_n = \frac{1}{6}[-x_{n-2} + 4x_{n-1} + 4x_{n+1} - x_{n+2}]$$

The noise amplification factor of this formula (Section 1.7) is 34/36.

Notice that supplying one isolated missing value is not the same as the usual dynamic process of imagining a stream of data passing through a digital filter. Nevertheless, we can look at what this formula does to any given frequency and thus look at its transfer function. To do so, we proceed as usual and substitute for the function x_n the complex frequency $e^{i\omega n}$. We find

$$H(\omega) = \frac{1}{3}[4\cos\omega - \cos 2\omega]$$

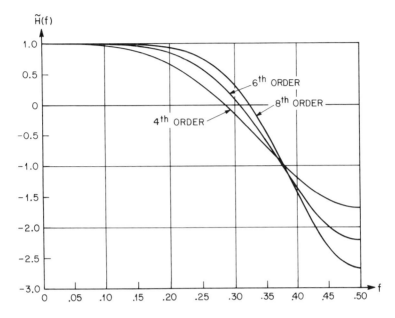

FIGURE 3.7-1 TRANSFER FUNCTION FOR MISSING DATA

as the transfer function, which should give the value 1 if the interpolated value is exactly what it should be. It is easy to see that we obtain the right answer for frequency 0. For higher frequencies, the value is not too good, particularly for very high frequencies. If the result seems to be strange, as shown in Fig. 3.7-1, then trying the limiting function in the Nyquist interval—namely, $1, -1, 1, -1, \ldots$—will reveal why the curves are what they are. Negative values on the graph of the transfer function imply a change in sign. The figure also shows the transfer function for sixth-and eighth-order differences equated to zero.

Again we see that the new way of looking at an old technique shows much more about what the formula does than is evident from mere inspection. We see the power of the frequency approach via the eigenfunctions of a linear, equally spaced sampled system. These formulas clearly show the danger of interpolating a missing value when the data is noisy—meaning that the data has numerous high frequencies.

Similar curves can be drawn for formulas that use the least-squares polynomial fit to the adjacent values as the basis for interpolating the missing value.

If we now turn to the problem of interpolating midpoint values in some data, then the simplest (linear) interpolation formula is

$$\frac{1}{2}[x_{n+1/2} + x_{n-1/2}]$$

Using four adjacent points and making the formula exact through cubics give the interpolation formula

$$\frac{1}{16}[-x_{n+3/2} + 9x_{n+1/2} + 9x_{n-1/2} - x_{n-3/2}]$$

The topic of interpolation will be treated in more detail in Section 7.1, where the transfer functions of these formulas appear.

Exercises

3.7-1   Derive the midpoint interpolation formulas given in the text.

3.7-2   In digital signal processing the data is often given at the midpoints and needs to be shifted $\Delta t/2$. We fitted 2, 4 points and found the interpolatory formula transfer functions for the midpoints. Derive the corresponding least-squares formulas for 4 and 6 points. Plot the transfer functions and compare with Fig. 7.1-1.

3.8   A CLASS OF NONRECURSIVE SMOOTHING FILTERS

Having looked at a large number of classical smoothing filters from the frequency approach, we are ready to attempt the *design* of a smoothing filter. At this point we will try only the simple class of symmetric filters,

$$y_n = ax_{n-2} + bx_{n-1} + cx_n + bx_{n+1} + ax_{n+2}$$

As usual, we start by asking what happens to a single frequency when we apply this smoothing filter; that is, we assume that

$$x_n = e^{i\omega n}$$

and examine the output. After putting this function in the equation and factoring out the exponential depending on n, we obtain the transfer function

$$H(\omega) = 2a \cos 2\omega + 2b \cos \omega + c$$

If the symmetry had not been assumed, there would have been some sine terms with imaginary coefficients.

How shall we choose the coefficients a, b, and c in our three-parameter family of filters? The answer depends, of course, on what we want to do, which frequencies we want to pass and which we want to stop. Suppose that, as usual, we want to pass low frequencies and stop high frequencies. Thus we want a lowpass filter. To convert a lowpass filter to a highpass filter, we need only compute $x_n - y_n$ as the new filter; what was stopped in the y_n is present in the x_n and is passed, whereas what is passed by the y_n is eliminated in the difference $x_n - y_n$.

We start by *arbitrarily* imposing two conditions on the transfer function; at the low frequency end we require

$$H(0) = 1 \qquad \text{exact at dc (lowest frequency)}$$

and at the upper end we require

$$H(\pi) = 0 \qquad \text{no highest frequency gets through}$$

These two conditions are equivalent to the pair of equations

$$H(0) = 2a + 2b + c = 1$$
$$H(\pi) = 2a - 2b + c = 0$$

From these two equations we get

$$b = \frac{1}{4}$$

$$c = \frac{1}{2} - 2a$$

and we are reduced to a one-parameter family of filters (see Fig. 3.8-1). The filters are

$$H(\omega) = 2a \cos 2\omega + \frac{1}{2} \cos \omega + \frac{1}{2} - 2a$$

or

$$H(\omega) = 4a[1 + \cos \omega \left[\cos \omega - \left(1 - \frac{1}{8a} \right) \right]$$

Note that $H(\omega)$ is a periodic function of ω and that it is an even function, however, as noted, because of sampling and the resulting aliasing, there is little

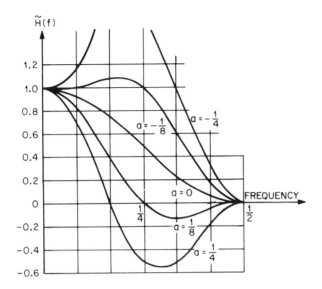

FIGURE 3.8-1   THE TRANSFER FUNCTION AS A FUNCTION OF THE PARA-
METER

meaning to the transfer function outside the Nyquist interval—the periodicity
of the mathematical expression reflects the aliasing rather than any analytic
property.

From Fig. 3.8-1 we can select a filter that approximately meets our needs.
Or we could impose one more condition on the transfer function $H(\omega)$ and
thus directly determine the filter. We will try this second approach in order to
illustrate the designing of very simple filters.

As a first example of the direct design of a filter, suppose that we require

$$H_1\left(\frac{\pi}{2}\right) = 1$$

which is equivalent to

$$H_1\left(\frac{\pi}{2}\right) = -2a + \frac{1}{2} - 2a = 1$$

This gives

$$a = -\frac{1}{8}, \qquad c = \frac{3}{4}$$

Therefore the filter is

$$y_n = \frac{1}{8}[-x_{n-2} + 2x_{n-1} + 6x_n + 2x_{n+1} - x_{n+2}]$$

and its transfer function is

$$H_1(\omega) = -\frac{\cos 2\omega}{4} + \frac{\cos \omega}{2} + \frac{3}{4}$$

As a second example of the design of a filter, suppose that we wish to balance the two halves of the filter by requiring

$$H_2\left(\frac{\pi}{2}\right) = \frac{1}{2}$$

Then we get, in turn,

$$H_2\left(\frac{\pi}{2}\right) = -2a + \frac{1}{2} - 2a = \frac{1}{2}$$

$$a = 0, \qquad c = \frac{1}{2}$$

Hence

$$y_n = \frac{1}{4}[x_{n-1} + 2x_n + x_{n+1}]$$

and

$$H_2 = \cos \frac{\omega}{2} + \frac{1}{2}$$

As a third example, suppose that we try to do as well as possible in the neighborhood of zero frequency. We already have both

$$H_3(0) = 1 \quad \text{and} \quad \frac{dH_3(0)}{d\omega} = 0$$

So we impose the further condition that

$$\frac{d^2H_3(0)}{d\omega^2} = 0 = -8a - \frac{1}{2}$$

or

$$a = -\frac{1}{16}$$

Thus we have

$$y_n = \frac{1}{16}[-x_{n-2} + 4x_{n-1} + 10x_n + 4x_{n+1} - x_{n+2}]$$

with

$$H_3(\omega) = -\frac{\cos 2\omega}{8} + \frac{\cos \omega}{2} + \frac{5}{8}$$

These filters are merely some of the simplest ones that we can design and are mainly for illustrative purposes. They are not intended as serious design problems.

3.9   AN EXAMPLE OF HOW A FILTER WORKS

Let us now examine *how* a filter does its job. For this purpose, we want the simplest example that we can get. We select two frequencies in the Nyquist interval, one at $f = \frac{1}{8}$ and the other at $f = \frac{3}{8}$. Next, we design the simple filter to exactly pass the first and stop the second; that is, we pick our filter form as

$$\tilde{H}(f) = 2a \cos 2\pi f + b$$

where we are now working in the rotation variable f instead of the earlier angular frequency variable ω. The two conditions on the filter that we are imposing are

$$\tilde{H}\left(\frac{1}{8}\right) = 1 \quad \text{and} \quad \tilde{H}\left(\frac{3}{8}\right) = 0$$

This leads to the filter with coefficients c_k

$$\frac{1}{2}[0.7, 1, 0.7]$$

where for clarity we have rounded the numbers to one significant digit ($1/\sqrt{2} = 0.7$).

We next construct the input signal consisting of the two frequencies, $\frac{1}{8}$ and $\frac{3}{8}$, of a rotation per sampling step. (See Table 3.9-1.) By placing the three

TABLE 3.9-1   Data to illustrate filtering

n	$\cos \dfrac{\pi n}{4}$	$\cos \dfrac{3\pi n}{4}$	Sum
0	1.0	1.0	2.0
1	0.7	−0.7	0.0
2	0.0	0.0	0.0
3	−0.7	0.7	0.0
4	−1.0	−1.0	−2.0
5	−0.7	0.7	0.0
6	0.0	0.0	0.0
7	0.7	−0.7	0.0
8	1.0	1.0	2.0

coefficients of our digital filter opposite *any* three consecutive values of the sum (column 4) we see that we get from the filter the entry in the first cosine column. Additional trials show that using the three filter coefficients on the first cosine (column 2) reproduces the entries, whereas on the second cosine (column 3) the result is always exactly zero (always, of course, within the roundoff level that we are using). This is a very simple, graphic illustration of how a filter does its job, how it can pass one frequency and stop another.

Exercise

3.9-1 Repeat this section, selecting a filter to pass zero frequency exactly and stop the frequency $f = \frac{1}{2}$.

3.10 SUMMARY

The purpose of this chapter was to show that the frequency approach provides a useful way to understand the various results of classical polynomial approximation. The examples were chosen for their ease of understanding and general familiarity rather than for their basic importance.

Before proceeding, it is necessary to develop some of the formal mathematics related to the Fourier series, which is the subject of the next two chapters. At each stage we shall include simple examples to illustrate the theory as it is developed, instead of going through a long stretch of formal mathematics before showing how and why it is relevant.

4

Fourier Series:
Continuous Case

4.1 NEED FOR THE THEORY

The preceding chapter showed that the transfer function of the typical digital filter takes the form of a sum of cosines of various integer frequencies (occasionally sines occur). It is not difficult to see that if our computed output values x_n refer to the midpoints between the samples, then for nonrecursive filters the transfer functions will be a sum of frequencies at the half odd integers (see Section 3.7). We will pursue this point further only in the exercises. (For instance, see Exercise 4.4-3.)

The filters considered were of three types. First, we had filters that ideally were either 0 or 1 at various places in the Nyquist interval: 0 where we wished to stop and 1 where we wanted to pass the frequencies of the signal. A common special case is the noise filter that typically passes the low frequencies and attenuates the high ones, the cutoff place depending on where we think the frequencies of the signal end. Secondly, we had differentiators where ideally we wished to approximate the function $H(\omega) = i\omega$ and were concerned with the ratio of the calculated to true values rather than simply the output of the filter. Finally, we had integrators where we wanted to approximate the function $1/i\omega$ and again were interested in ratios. Clearly, other types of filters could occur, and we need to be able to cope with any reasonably shaped transfer function.

In order to design filters, rather than merely evaluate them, we must go fairly directly from the proposed transfer function to the expansion in terms of the trigonometric functions. It will then be easy to go to the actual coefficients c_k of the digital filter. Fundamental to this design process, therefore, is the expansion of an arbitrary function in terms of the trigonometric functions. This is *the theory of Fourier series*, the subject of this and the next chapter.

The relationship of formal mathematics to the real world is ambiguous. Apparently, in the early history of mathematics the mathematical abstractions of integers, fractions, points, lines, and planes were fairly directly based on experience in the physical world. However, much of modern mathematics seems to have its sources more in the internal needs of mathematics and in aesthetics rather than in the needs of the physical world. Since we are interested mainly in *using* mathematics, we are obliged in our turn to be ambiguous with respect to mathematical rigor. Those who believe that mathematical rigor justifies the use of mathematics in applications are referred to Lighthill [Li] and Papoulis [P] for rigor; those who believe that it is the usefulness in practice that justifies the mathematics are referred to the rest of this book. We adopt the attitude that useful mathematics can be mathematically justified, even if it is necessary to alter classical definitions and postulates of mathematics (recall the recent change from "function" to "generalized function" to justify the widespread use of the Dirac delta function). Furthermore, since we are interested in the anatomy of the mathematics, we shall ignore many of the mathematically pathological cases. The fact that we are dealing with samples of a physical function implies that we are trying to understand a reasonable situation.

In short, for our purposes, the justification of the various mathematical models and their rigor rests more on their usefulness in the real world than on the internal aesthetics of mathematics.

4.2 ORTHOGONALITY

Although we shall want to expand $H(\omega)$ or $\tilde{H}(f)$, where ω and f are, respectively, the independent variable, it is convenient to give the theory of Fourier series in terms of the independent variable t. This step will cause some confusion later, but we hope to label the places where it may occur. Meanwhile, our notation will be closer to other texts in Fourier series. Moreover, the Fourier series has many uses beyond digital filter theory, so that the neutral notation t enables the reader to use what he learns here in quite different fields.

The first concept needed is that of *orthogonality*. Two functions $g_1(t)$ and

$g_2(t)$ (neither identically zero) are said to be *orthogonal* with respect to a weight function $K(t) \geq 0$, in an interval $a \leq t \leq b$, if

$$\int_a^b K(t)g_1(t)g_2(t)\, dt = 0$$

This concept is a large extension of the idea of orthogonal lines in n-dimensional space. To see this point, consider two n-dimensional vectors $U = \{u_1, u_2, \ldots, u_n\}$ and $V = \{v_1, v_2, \ldots, v_n\}$. In forming the sum of the two vectors, we add components term by term. The sum is the third side of the triangle with U and V being the other two sides. If it is to be a right triangle with U and V as the legs, then we apply the Pythagorean Theorem and assert that

$$(U + V)^2 = U^2 + V^2$$

Multiply out this vector equation and cancel the square terms from both sides, leaving the sum of the cross products (ignoring a factor of two)

$$\sum_{k=0}^n u_k v_k = 0$$

By suitably regarding this sum while letting the dimension n become infinite, we are led to the corresponding integral

$$\int_0^1 u(k)v(k)\, dk$$

The kernel $K(t)$ of the integrand in the definition is a nonnegative weighting factor on the various components and adds no new complications. Notice, however, that the integral represents a noncountable number of dimensions.

A set of functions $g_n(t)$, $n = 0, 1, 2, \ldots$, is said to be orthogonal if

$$\int_a^b K(t)g_m(t)g_n(t)\, dt = \begin{cases} 0, & \text{when } m \neq n \\ \lambda_n^2 & \text{when } m = n \end{cases} \qquad [K(t) \geq 0]$$

Of course, when $m = n$, the integrand is nonnegative and the integral must be a positive number. If $\lambda_n = 1$ for all n, then they are said to be *an orthonormal family* of functions. It is easy to convert an orthogonal set to an orthonormal set by merely dividing the nth function by the corresponding λ_n. Orthonormality is a convenient property in theoretical work because it eliminates the appearance of the λ_n. However, in practice, it is usually not done because it introduces awkward numerical factors.

Perhaps the best-known set of orthogonal functions is the Fourier set

$$1, \cos t, \cos 2t, \cos 3t, \ldots$$

$$\sin t, \sin 2t, \sin 3t, \ldots$$

over the interval $0 \leq t \leq 2\pi$ (or else $-\pi \leq t \leq \pi$). To show that this set is orthogonal, we need to derive the following three integrals:

$$\int_0^{2\pi} \cos mt \cos nt \, dt = \begin{cases} 0, & m \neq n \\ \pi, & m = n \neq 0 \\ 2\pi, & m = n = 0 \end{cases}$$

$$\int_0^{2\pi} \cos mt \sin nt \, dt = 0$$

$$\int_0^{2\pi} \sin mt \sin nt \, dt = \begin{cases} 0, & m \neq n \\ \pi, & m = n \neq 0 \\ 0, & m = n = 0 \end{cases}$$

To derive the first, we use the trigonometric identity

$$\cos mt \cos nt = \frac{1}{2}[\cos (m + n)t + \cos (m - n)t]$$

Integrating, we get for $m \neq n$

$$\frac{1}{2}\left[\frac{\sin (m + n)t}{(m + n)} + \frac{\sin (m - n)t}{(m - n)}\right]$$

When the limits 2π and 0 are inserted, this is clearly zero. When $m = n \neq 0$, we use the identity

$$\cos nt \cos nt = \cos^2 nt = \frac{1}{2}[1 + \cos 2nt]$$

Integrating, we get

$$\frac{1}{2}\left[t + \frac{\sin 2nt}{2n}\right]$$

which upon insertion of the limits gives the value π. Finally, when $m = n = 0$, the integrand clearly is the constant 1, and the integral therefore is 2π. The other orthogonality relations can be similarly verified by using the corresponding trigonometric identities.

Example. To normalize the trigonometric functions, we need to divide the sines and cosines by

$$\sqrt{\pi}$$

and the constant term 1 by

$$\sqrt{2\pi}$$

Exercises

4.2-1   Show that the cosines are orthogonal over the interval $0 \le x \le \pi$.

4.2-2   Show that the sines are orthogonal over the interval $0 \le x \le \pi$.

4.2-3   Complete the proof that the Fourier set of functions is orthogonal.

4.2-4   Show the orthogonality applies also for any interval of length 2π—in particular, for $-\pi \le x \le \pi$.

4.3   FORMAL EXPANSIONS

Given a function $g(t)$, $0 \le t \le 2\pi$, we assume that $g(t)$ has the formal expansion

$$g(t) = \frac{a_0}{2} + \sum_{k=1}^{\infty} [a_k \cos kt + b_k \sin kt]$$

The reason for the $a_0/2$ term will become clear shortly. To obtain an expression for the *coefficients* a_k, we multiply the equation on both sides by $\cos mt$ and integrate $0 \le t \le 2\pi$. As a result of the orthogonality, we obtain

$$\int_0^{2\pi} g(t) \cos mt \; dt = \begin{cases} \pi a_m, & m \ne 0 \\ \pi a_0, & m = 0 \end{cases}$$

To get the b_k, we multiply by $\sin mt$ in place of $\cos mt$ and integrate

$$\int_0^{2\pi} g(t) \sin mt \; dt = \pi b_m \qquad (m \ne 0)$$

Thus the coefficients of the assumed expansion are given by the formulas

$$a_m = \frac{1}{\pi} \int_0^{2\pi} g(t) \cos mt \, dt \qquad (m = 0, 1, 2, \ldots)$$

$$b_m = \frac{1}{\pi} \int_0^{2\pi} g(t) \sin mt \, dt \qquad (m = 1, 2, 3, \ldots)$$

These a_m and b_m are called the *Fourier coefficients* of the expansion of $g(t)$.

Notice that when the value of t falls outside the original range $0 \le t \le 2\pi$, the function $g(t)$ is defined by the expansion to be periodic; thus we have $g(2\pi + t) = g(t)$ for all t.

Example. To illustrate, suppose that $g(t) = t$ and that we use the interval $-\pi \le t \le \pi$. Since the integrand is odd and the range of integration is symmetric about $t = 0$, we have

$$a_m = \frac{1}{\pi} \int_{-\pi}^{\pi} t \cos mt \, dt = 0$$

For the b_m, we have

$$b_m = \frac{1}{\pi} \int_{-\pi}^{\pi} t \sin mt \, dt$$

Integrating by parts,

$$b_m = \frac{1}{\pi} \left[\frac{t(-\cos mt)}{m} \Big|_{-\pi}^{\pi} + \frac{1}{m} \int_{-\pi}^{\pi} \cos mt \, dt \right]$$

$$= \frac{1}{\pi} \left[\frac{\pi}{m} 2(-1)^{m+1} + \frac{1}{m} \frac{\sin mt}{m} \Big|_{-\pi}^{\pi} \right]$$

$$= \frac{2}{m} (-1)^{m+1}$$

Thus we have the formal expansion

$$t = 2 \left[\sin t - \frac{\sin 2t}{2} + \frac{\sin 3t}{3} - \frac{\sin 4t}{4} + \ldots \right]$$

Figure 4.3-1 shows the first few partial sums (labeled S_N) as approximations to the function $g(t) = t$. Notice the effect of the periodicity at the ends of the interval, $-\pi \le t \le \pi$,

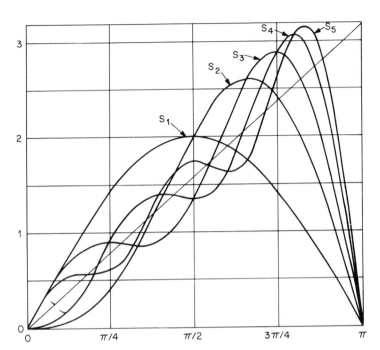

FIGURE 4.3-1   PARTIAL SUMS OF $S_N = \sum\limits_{n=1}^{N} (-1)^n \dfrac{\sin{(nt)}}{n}$

Example. As a second illustration of the expansion of a given function into a formal Fourier series, consider the "rectangular pulse"

$$g(t) = \begin{cases} \dfrac{1}{2}, & 0 \le t \le \pi \\[2mm] -\dfrac{1}{2}, & -\pi \le t \le 0 \end{cases}$$

Since

$$g(-t) = -g(t)$$

there will be no cosine terms in the final expansion. The coefficients of the sine terms are given by

$$b_k = \frac{1}{\pi} \int_{-\pi}^{\pi} g(t) \sin{kt}\; dt$$

$$= \frac{1}{\pi} \int_{0}^{\pi} \sin{kt}\; dt$$

Doing the integration we have

$$b_k = \frac{1}{\pi} \frac{[1 + (-1)^{k+1}]}{k} = \begin{cases} \dfrac{2}{\pi k}, & k \text{ odd} \\ 0, & k \text{ even} \end{cases}$$

Thus we have the formal expansion

$$g(t) = \frac{2}{\pi} \left[\sin x + \frac{1}{3} \sin 3x + \frac{1}{5} \sin 5x + \dots \right]$$

$$= \frac{2}{\pi} \sum_{k=0}^{\infty} \frac{1}{2k+1} \sin (2k+1)t$$

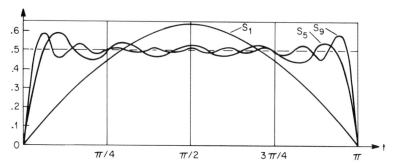

FIGURE 4.3-2   PARTIAL SUMS S_1, S_5, S_9 FOR RECTANGULAR PULSE

Fig 4.3-2 gives the graphs of the partial sums (labeled S_N) for 1, 5, and 9 terms of the series, for the interval $0 \le t < \pi$. For $-\pi < t \le 0$, the curves are the negatives of those shown. The figure indicates the quality of the approximation. We will return to these examples later.

Exercises

4.3-1   Write out the formulas for a Fourier expansion over the interval $0 \le x \le L$.

4.3-2   Show that the function

$$g(x) = \begin{cases} 1, & 0 < x < \dfrac{L}{2} \\ -1, & \dfrac{L}{2} < x < L \end{cases}$$

has the form

$$g(x) = \frac{4}{\pi} \sum_{m=1}^{\infty} \frac{1}{2m-1} \sin \left[\frac{(2m-1)2\pi x}{L} \right]$$

4.3-3   Show that the function

$$g(x) = \begin{cases} \dfrac{2x}{L}, & 0 < x < \dfrac{1}{2} \\[2mm] \dfrac{2 - 2x}{L}, & \dfrac{L}{2} < x < L \end{cases}$$

has the formal Fourier expansion

$$g(x) = \frac{1}{2} - \frac{4}{\pi^2} \sum_{m=1}^{\infty} \frac{1}{(2m-1)^2} \cos\left[\frac{(2m-1)2\pi x}{L}\right]$$

4.3-4   Find the expansion of $g(x) = |\sin x|$ for $0 \le x \le 2\pi$.

4.3-5   Find the expansion of the function $g(t) = \sin x$ for $0 \le x \le \pi$ and $g(x) = 0$ for $\pi \le x \le 2\pi$.

4.4   ODD AND EVEN FUNCTIONS

Odd and even functions occur so often that their Fourier series are worth special examination. The fact that an arbitrary function $g(t)$ can be written as the sum of an odd function plus an even function simply by writing

$$g(t) = \frac{[g(t) - g(-t)]}{2} + \frac{[g(t) + g(-t)]}{2}$$

makes this point particularly true.

For an odd function, we see that $a_k = 0$ for all k, whereas

$$b_k = \frac{2}{\pi} \int_0^{\pi} g(t) \sin kt \, dt$$

Our differentiator filters (Section 3.5) were typical odd functions.

For an even function, we have

$$a_k = \frac{2}{\pi} \int_0^{\pi} g(t) \cos kt \, dt$$

and all the $b_k = 0$. The typical smoothing filter (Sections 3.2, 3.3, 3.6, and 3.7) is an even function.

By using additional suitable symmetries about $t = \pi/2$, we can obtain Fourier series with, say, only odd-indexed coefficients or only even-indexed coefficients. Other regular patterns of nonzero coefficients can also be obtained by suitably introducing proper odd and even symmetries into the definition of the function.

Exercises

4.4-1   Expand $g(t) = e^{-a|t|}$, $|t| \leq \pi$.

4.4-2   Expand the function $g(t) = t^2$, $t \geq 0$, and $g(-t) = -g(t)$, $t \leq 0$, for the range $-L \leq t \leq L$.

4.4-3   For an even function in the interval $-2\pi \leq x \leq 2\pi$ and $g(2\pi - x) = g(x)$ show that $b_n = 0$ and

$$a_n = \begin{cases} \dfrac{2}{\pi} \displaystyle\int_0^\pi g(x) \cos \dfrac{n}{2} x \, dx, & n \text{ odd} \\[4mm] 0, & n \text{ even} \end{cases}$$

4.5   FOURIER SERIES AND LEAST SQUARES

The Fourier series expansion is closely connected with the least-squares approximation of a function; indeed, we will show that the Fourier coefficients give the least-squares fit to the function. To do so, we set up the usual "sum of the squares of the residuals" for measuring the approximation of the function $g(t)$ by $g_N(t)$, where

$$g(t) = \frac{a_0}{2} + \sum_{k=1}^{\infty} a_k \cos kt + \sum_{k=1}^{\infty} b_k \sin kt$$

and
$$g_N(t) = \frac{A_0}{2} + \sum_{k=1}^{N} A_k \cos kt + \sum_{k=1}^{N} B_k \sin kt$$

Note the use of lowercase and capital letters, where A_k and B_k are arbitrary coefficients selected for the fit. Thus

$$S^2 = \frac{1}{\pi} \int_{-\pi}^{\pi} [g(t) - g_N(t)]^2 \, dt \geq 0$$

is the relevant sum of squares. The formal Fourier coefficients are given by the usual formulas (note the lowercase letters here)

$$a_k = \frac{1}{\pi} \int_{-\pi}^{\pi} g(t) \cos kt \, dt$$

$$b_k = \frac{1}{\pi} \int_{-\pi}^{\pi} g(t) \sin kt \, dt$$

But subtracting the two expansions term by term (where they match), squaring, and using the orthogonality of 1, cos t, sin t, cos $2t$, . . . , gives

$$S^2 = \frac{[A_0 - a_0]^2}{2} + \sum_{k=1}^{\infty} [\{A_k - a_k\}^2 + \{B_k - b_k\}^2] \geq 0$$

In these sums the A_k and B_k for $k > N$ are to be taken as zero. Since we want a least-squares fit, S^2 will clearly be a minimum if an only if *for all k* less than or equal to N

$$A_k = a_k \qquad B_k = b_k$$

For finite N we have therefore proved that the formal Fourier expansion is the least-squares fit.

If the equality

$$\frac{1}{\pi} \int_{-\pi}^{\pi} g^2(t) \, dt = \frac{a_0^2}{2} + \sum_{k=1}^{\infty} [a_k^2 + b_k^2]$$

occurs, it is called *Parseval's equality.*

The inequality which occurs for finite N is called *Bessel's inequality*

$$\frac{1}{\pi} \int_{-\pi}^{\pi} g^2(t) \, dt \geq \left\{ \frac{a_0^2}{2} + \sum_{k=1}^{N} (a_k^2 + b_k^2) \right\}$$

Bessel's inequality is very useful in estimating the error of the fit as we gradually increase the number of coefficients being computed. When rearranged, this equation says that the sum of the squares of the residuals of the fit (approximation) is the difference between the integral of the square of the function (with a factor of $1/\pi$) and the sum of the squares of the coefficients that have been computed. (Remember to take $a_0^2/2$.)

It is clear from Bessel's inequality that the series

$$\frac{a_0^2}{2} + \sum_{k=1}^{\infty} \{a_k^2 + b_k^2\}$$

is bounded above and therefore converges, provided that the functions $g(t)$ and $g^2(t)$ are integrable. Thus the Fourier coefficients of the expansion

$$a_k = \frac{1}{\pi} \int_{-\pi}^{\pi} g(t) \cos kt \, dt \longrightarrow 0$$

$$b_k = \frac{1}{\pi} \int_{-\pi}^{\pi} g(t) \sin kt \, dt \longrightarrow 0$$

as k approaches infinity. As a corollary to this result (which will be needed later), we have, using the addition formula for sines and simple bounds,

$$\frac{1}{\pi} \int_{-\pi}^{\pi} g(t) \sin \left(k + \frac{1}{2} \right) t \, dt \longrightarrow 0$$

as k approaches infinity.

Exercises

4.5-1   Prove the corollary.

4.5-2   Apply Bessel's inequality to the first few terms of the first example of Section 4.3.

4.5-3   Estimate (one and a half significant figures) the least-squares error for $N = 1, 5$, and 9 for the second example of Section 4.3.

4.6   CLASS OF FUNCTIONS AND RATE OF CONVERGENCE

Under the assumption that $g(t)$ and its square are both integrable, we have found that the Fourier coefficients a_k and b_k both approach zero as k approaches infinity. This result does not prove that the series converges, much less that, if it does converge, it will approach the original function $g(t)$. So let us investigate the rate of convergence and then examine what the series approaches.

For our applications, which class of functions should be considered? In almost all applied problems we need at worst a function consisting of a finite number of pieces (adjacent intervals) such that in each interval all the required derivatives exist (we may suppose that it is analytic in each piece if necessary). Thus we use piecewise analytic functions. This piecewise property means that if we wish to be very careful, we must speak of left-hand and right-hand derivatives at the ends of the intervals. However, we shall assume the reader's awareness. This class of functions allows jumps, corners, and similiar features—jumps in the function if the function is discontinuous, corners in the function if the first derivative is discontinuous, and sudden changes in the curvature if the second derivative is discontinuous, and so on.

Given a function of this class, what are its Fourier coefficients? We have for J pieces (intervals)

$$a_k = \frac{1}{\pi} \int_{-\pi}^{\pi} g(t) \cos kt \, dt$$

$$= \frac{1}{\pi} \sum_{j=1}^{J} \int_{t_{j-1}}^{t_j} g(t) \cos kt \, dt$$

where $t_0 = -\pi$, $t_J = \pi$, and the other t_j's mark the ends of the pieces where the breaks in the function occur. A similar formula applies for the b_k.

Next, integrating by parts, we have

$$a_k = \frac{1}{\pi} \sum_{j=1}^{J} \left[g(t) \left(\frac{\sin kt}{k} \right) \Big|_{t_{j-1}}^{t_j} - \int_{t_{j-1}}^{t_j} g'(k) \left(\frac{\sin kt \, dt}{k} \right) \right]$$

If the function is continuous [remember that this includes the condition that $g(-\pi) = g(\pi)$], then the integrated part will cancel out for all k. However, if the function is not continuous, it will not always cancel out, and the coefficients a_k will, in general, be of order $1/k$. If cancellation does occur, we can integrate by parts again

$$a_k = \frac{1}{\pi} \sum_{j=1}^{J} \left[g'(t) \left(\frac{\cos kt}{k^2} \right) \Big|_{t_{j-1}}^{t_j} - \int_{t_{j-1}}^{t_j} \frac{g''(t) \cos kt}{k^2} \, dt \right]$$

This time it is the continuity of the first derivative that is necessary if the integrated part is to cancel out. Continuing in this way, we find that without continuity of the function the coefficients of the Fourier expansion have some terms like $1/k$, that without continuity of the first derivative the coefficients are like $1/k^2$, that without continuity of the second derivative they are like $1/k^3$, and so forth. Similar statements are true for the b_k coefficients.

Once the coefficients are of order $1/k^2$, then, since the trigonometric functions are not greater than 1 in size, we have the convergence of the Fourier series. Only in the case of a discontinuous function need we look more closely.

4.7   CONVERGENCE AT A POINT OF CONTINUITY

In this section we consider the convergence of the formal Fourier series. Specifically, we ask: Does the series converge at a point *inside* one of the intervals $t_j \leq t \leq t_{j+1}$, *and* if so, does it approach the function at the point of continuity? In the next section we will discuss the convergence at points of discontinuity.

We have, for the sum of the frequencies up to N, the partial sum

$$g_N(t) = \frac{a_0}{2} + \sum_{k=1}^{N} [a_k \cos kt + b_k \sin kt]$$

where, of course, the Fourier coefficients are given by

$$a_k = \frac{1}{\pi} \int_{-\pi}^{\pi} g(s) \cos ks \, ds$$

$$b_k = \frac{1}{\pi} \int_{-\pi}^{\pi} g(s) \sin ks \, ds$$

The dummy variable of integration s has been used in order to avoid confusion. Eliminating the coefficients a_k and b_k in the partial sum, we have, on interchanging the finite summation and integration operations,

$$g_N(t) = \frac{1}{\pi} \int_{-\pi}^{\pi} g(s) \left[\frac{1}{2} + \sum_{k=1}^{N} \{\cos ks \cos kt + \sin ks \sin kt\} \right] ds$$

or $$g_N(t) = \frac{1}{\pi} \int_{-\pi}^{\pi} g(s) \left[\frac{1}{2} + \sum_{k=1}^{N} \cos k(s - t) \right] ds$$

For convenience, we assume that $g(t)$ is periodic; therefore by setting

$$s - t = u$$

we shift the coordinate system and place the point t (where we are examining the convergence question) in the middle of the interval of integration in the variable u. The result is

$$g_N(t) = \frac{1}{\pi} \int_{-\pi}^{\pi} g(t + u) \left[\frac{1}{2} + \sum_{k=1}^{N} \cos ku \right] du$$

To sum the expression in the square bracket

$$s_N = \frac{1}{2} + \sum_{k=1}^{N} \cos ku$$

we multiply through by $\sin u/2$ and apply some elementary trigonometry to get

$$\left(\sin \frac{u}{2}\right) s_N = \frac{1}{2}\left[\sin \frac{u}{2} + \left(\sin \frac{3u}{2} - \sin \frac{u}{2}\right) + \left(\sin \frac{5u}{2} - \sin \frac{3u}{2}\right)\right.$$

$$\cdots$$

$$\left. + \left(\sin \frac{(2N+1)u}{2} - \sin \frac{(2N-1)u}{2}\right)\right]$$

$$= \frac{1}{2} \sin \frac{(2N+1)u}{2}$$

Because of the cancellation of terms, only the last one remains; and we have for the sum of the cosines

$$s_N = \frac{\sin (2N+1)u/2}{2 \sin u/2}$$

Therefore the partial sum of the Fourier series is

$$g_N(t) = \frac{1}{\pi} \int_{-\pi}^{\pi} g(t+u) \frac{\sin \{(2N+1)u/2\}}{\sin u/2} \frac{du}{2}$$

To prove the convergence of the partial sums to the function $g(t)$, we need an expression for the difference

$$g_N(t) - g(t)$$

To get it, we note that integrating both expressions for s_N from $-\pi$ to π gives

$$1 = \frac{1}{\pi} \int_{-\pi}^{\pi} \left[\frac{\sin (2N+1)u/2}{\sin u/2}\right] \frac{du}{2}$$

Multiplying by $g(t)$ and using the fact that the integration is with respect to u, we get the result

$$g(t) = \frac{1}{\pi} \int_{-\pi}^{\pi} [g(t)] \left[\frac{\sin (2N+1)u/2}{\sin u/2}\right] \frac{du}{2}$$

Subtraction of this result from the formula for the partial sum provides the needed expression

$$g_N(t) - g(t) = \frac{1}{\pi} \int_{-\pi}^{\pi} [g(t + u) - g(t)] \left[\frac{\sin (2N + 1)u/2}{\sin u/2} \right] \frac{du}{2}$$

To prove the convergence of the partial sum $g_N(t)$ to $g(t)$ at the point t, it is necessary to show that this difference approaches zero as N increases. To do so, we simply insert a factor u in both the numerator and the denominator and then rearrange the expression

$$g_N(t) - g(t) = \frac{1}{\pi} \int_{-\pi}^{\pi} \left[\frac{\{g(t + u) - g(t)\}}{u} \right] \left[\frac{\sin (2N + 1)u/2}{(\sin u/2)/(u/2)} \right] du$$

In this form the first square bracket approaches the derivative of $g(t)$ as u approaches 0, so that this term gives no trouble. The denominator of the second square bracket has the limit 1 as u approaches 0 and is well behaved elsewhere in the range of integration. Thus the function

$$\phi(u) = \frac{g(t + u) - g(t)}{2 \sin u/2}$$

is well behaved; and if it is integrable square, we can apply the earlier corollary to the integral

$$\frac{1}{\pi} \int_{-\pi}^{\pi} \phi(u) \sin \left[\frac{(2N + 1)u}{2} \right] du$$

Clearly, as N approaches infinity, the difference between the partial sum and the function approaches zero at any point at which a two-sided derivative exists. Thus we have convergence to the function.

THEOREM. *The formal Fourier series converges to the function at point where the derivative exists* [*provided that* $\phi^2(u)$ *is integrable*].

4.8 CONVERGENCE AT A POINT OF DISCOUNTINUITY

At a point of discontinuity t_j, that is at the end of some interval being used, we do not have a suitable derivative and the preceding argument breaks down.

The first obvious step is to break the integral into two halves, since the function is different on the two sides of a point t_j. As a result of the shift in the coordinate system, the point t_j is now at 0. We can write

$$g_N(t) = \frac{1}{\pi} \int_{-\pi}^{0} + \frac{1}{\pi} \int_{0}^{\pi}$$

In the first integral we replace u by $-u$ and then combine both integrals:

$$g_N(t) = \frac{1}{\pi} \int_{-\pi}^{\pi} [g(t_j + u) + g(t_j - u)]\frac{\sin (2N + 1)u/2}{\sin u/2}\frac{du}{2}$$

In the earlier proof, when at a point of continuity, we subtracted out $g(t)$. What must we do now? Clearly, we need something for both halves—namely,

$$\frac{g(t_j +) + g(t_j -)}{2}$$

(the $+$ and $-$ signs mean limit from the right and left, respectively). This is the average of the two limiting values (the factor one-half occurs because the range of integration is now one-half of what it was previously). Using this value, we obtain an expression for the difference

$$g_N(t_j) = -\left[\frac{g(t_j +) + g(t_j -)}{2}\right]$$

which will approach zero as before (although we need an extra argument to prove that the corollary can be suitably modified for the half-interval). Thus we have

THEOREM.   *At a point of discontinuity the formal Fourier series converges to the average of the two limiting values of the function* [(*provided that* $\phi^2(u)$ *is integrable*].

We have seen this effect in the two examples of Section 4.3.

It is probably worth pointing out that the *convergence* of a Fourier series at a point is a local property, whereas the *rate of convergence* is a global property.

Exercise

4.8-1   Fill in the details of this section.

5

Further Results
in Fourier Series

5.1 INTRODUCTION

Of the many results in the theory of Fourier series, only a comparative few need be examined. One is the famous Gibbs phenomenon. Chapter 4 showed that, for reasonable functions, the formal Fourier series converges to the function at points of continuity and converges to the average of the two limits at a point of discontinuity. However, each term in the Fourier series is a continuous function; consequently, the theorem "a *uniformly* convergent series of continuous functions converges to a continuous function" implies that at a point of discontinuity the convergence of the Fourier series cannot be uniform—something peculiar must occur. This occurrence is the *Gibbs phenomenon*. For those unfamiliar with the idea of uniform convergence, the subject is discussed briefly in Appendix 5.A of this chapter.

We shall also develop the complex form of the Fourier series, which is less closely related to the direct measurement of quantities that occur in real problems but is much more suitable for mathematical manipulation. Once this new form is mastered, it provides additional insight to both the theory and the underlying physical phenomena.

Another idea is also needed, the so-called convolution theorems. Such theorems, although often ignored in a general treatment of the Fourier series, are essential to filter theory.

5.2   GIBBS PHENOMENON

The purpose of this section is to examine the Gibbs phenomenon, named for J. Willard Gibbs (who first publicized this effect but was not the first to publish it). The second example of Section 4.3 which was the expansion of the rectangular pulse function, is typical of a discontinuous function. This function was defined as

$$g(t) = \begin{cases} -\dfrac{1}{2}, & -\pi < t < 0 \\[2ex] \dfrac{1}{2} & 0 < t < \pi \end{cases}$$

and has a discontinuity of size 1 at $t = 0$. The formal Fourier expansion was found to be

$$g(t) = \frac{2}{\pi} \sum_{k=0}^{\infty} \frac{\sin (2k + 1)t}{2k + 1}$$

We can write the truncated series of partial sums as

$$g_N(t) = \frac{2}{\pi} \sum_{k=0}^{N} \left[\int_0^t \cos (2k + 1)u \, du \right] = \frac{2}{\pi} \int_0^t \left[\sum_{k=0}^{N} \cos (2k + 1)u \right] du$$

To get the sum of the cosines s_N; we multiply, as before, by the sine of the angle of one-half the difference of the angles in successive terms; the result is (compare with Section 4.7)

$$s_N \sin u = \sum_{k=0}^{N} \sin u \cos (2k + 1)u = \frac{1}{2} \sin 2(N + 1)u$$

Thus we have the partial sums as integral

$$g_N(t) = \frac{1}{\pi} \int_0^t \frac{\sin 2(N + 1)u}{\sin u} \, du$$

Figure 5.2-1 illustrates the partial sum behavior as a function of t in the case of five terms and with the independent variable $x = t/\pi$. We see an overshoot of the curve of the partial sums. To find where the peaks and valleys occur, we differentiate the function and set it equal to zero. This procedure is the same as equating the integrand to zero. Clearly, the first maximum

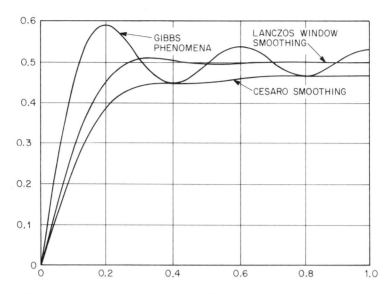

FIGURE 5.2-1 GIBBS PHENOMENA WITH WINDOWS: PLAIN, LANCZOS, AND CESARO WINDOWS

occurs at $t = \pi/2(N + 1)$. The value of the partial sum at this point is

$$g_N\left[\frac{\pi}{2(N + 1)}\right] = \frac{1}{\pi}\int_0^{\pi/2(N+1)} \frac{\sin 2(N + 1)u}{\sin u}\,du$$

Our problem is to estimate this value as N approaches infinity. Using the change of the dummy variable of integration $v = 2(N + 1)u$ gives (when we insert extra v terms)

$$g_N\left[\frac{\pi}{2(N + 1)}\right] = \frac{1}{\pi}\int_0^{\pi} \left(\frac{\sin v}{v}\right)\left[\frac{\{v/2(N + 1)\}}{\sin\{v/2(N + 1)\}}\right]\,dv$$

Obviously, since $(\sin \theta)/\theta$ approaches 1 as θ approaches 0, the second square bracket approaches 1 as N gets larger and larger. Thus in the limit we have, for the value of the first overshoot,

$$g_\infty(0) = \frac{1}{\pi}\int_0^{\pi} \frac{\sin v}{v}\,dv$$

This is the standard sine integral function. From tables, we get the value for the first maximum

$$\frac{1.17898}{2} = 0.58949 = 0.5 + 0.08949$$

The next local minimum has the corresponding limiting value of

$$0.45142 = 0.5 - 0.04858$$

These values lead (when N is large) to the values of approximately 9 % overshoot and 5 % undershoot for the unit jump (discontinuity) in the function.

This Gibbs phenomenon occurs *whenever* we truncate a Fourier series. It will be of great importance when discussing the design of filters because, for practical purposes, we are forced to truncate any Fourier series we generate.

5.3   LANCZOS SMOOTHING: THE SIGMA FACTORS

Cornelius Lanczos observed that the ripple in the sum of the truncated series has the period of either the first term neglected or the last term kept. In either case, he argued, smoothing the partial sum by integrating (averaging) over this period would remove the main effects of the ripple.

We apply this idea to the general truncated Fourier series

$$g_N(t) = \frac{a_0}{2} + \sum_{k=1}^{N} [a_k \cos kt + b_k \sin kt]$$

For the smoothed value, we take the *average* over the interval of length $2\pi/N$ centered at t. (Here we pick N as the order of the last term kept so that the reader can watch the last term disappear, but, in practice, we pick N as the order of the first term neglected.) We have, therefore, the smoothed value $h_N(t)$ as the average of the $g_N(t)$

$$h_N(t) = \frac{N}{2\pi} \int_{t-(\pi/N)}^{t+(\pi/N)} g_N(s)\, ds$$

Working out the details, we have

$$h_N(t) = \left(\frac{a_0}{2}\right)\left(\frac{N}{2\pi}\right) \int_{t-(\pi/N)}^{t+(\pi/N)} ds$$

$$+ \frac{N}{2\pi} \sum_{k=1}^{N} \left[a_k \int_{t-(\pi/N)}^{t+(\pi/N)} \cos ks\, ds + b_k \int_{t-(\pi/N)}^{t+(\pi/N)} \sin ks\, ds \right]$$

$$h_N(t) = \frac{a_0}{2} + \frac{N}{2\pi} \sum_{k=1}^{N} \left[a_k \left\{ \frac{\sin k[t + (\pi/N)] - \sin k[t - (\pi/N)]}{k} \right\} \right.$$

$$\left. - b_k \left\{ \frac{\cos k[t + (\pi/N)] - \cos k[t - (\pi/N)]}{k} \right\} \right]$$

Using the trigonometric formula for the difference of two sines and the formula for the difference of two cosines, we finally obtain

$$h_N(t) = \frac{a_0}{2} + \sum_{k=1}^{N} \sigma(N, k)[a_k \cos kt + b_k \sin kt]$$

where $\sigma(N, k)$ are the so-called *sigma factors*

$$\sigma(N, k) = \frac{\sin \pi k/N}{\pi k/N}$$

Thus the smoothed Fourier series is the original Fourier series with its coefficients multiplied by the corresponding sigma factors.

This result is not surprising; since the smoothing operation was linear, some kind of multiplying factor had to appear. When $k = N$, the sigma factor is zero, and it does not matter whether the last term kept is called N or the first term neglected is called N: the effect is that the term of frequency N has the multiplier equal to zero.

The formula for $h_N(t)$ is the average of $g_N(t)$ for a symmetric interval about t and of width $2\pi/N$. We may view this smoothing operation as if looking at the original function $g(t)$ through a narrow, translucent, rectangular window of width $2\pi/N$. The function $h_N(t)$ that we see is the average light intensity from the original function $g_N(t)$, which is imagined as being brighter when the function $g_N(t)$ is higher and less bright when the function is smaller. It is easily seen that the resultant function $h_N(t)$, the Lanczos' smoothed curve, has the high-frequency ripples somewhat smoothed out as shown by Fig. 5.2-1.

In selecting our smoothing interval, we adjusted it to the number of terms being kept in the Fourier series; that is, we had the same N in both the number of terms and in the smoothing interval or, what is the same thing, in the sigma factors. An examination of the process of deriving the sigma factors shows that if we do not make these two the same (or differing by 1), then we will still obtain sigma factors that are related to the smoothing interval selected. A study of the sigma factors shows that in this situation the series will not necessarily end with the Nth sigma factor that has the value 0 but can continue with additional sigma factors that are not zero. Viewed as a function of k, the sigma factors when k is greater than N have a decaying ripple of period $2N$ and are slowly alternating in sign.

The theory of Fourier series has another smoothing formula that is widely used (mainly in mathematical circles)—namely, the averaging of the successive partial sums $g_N(t)$. The process is called *Fejer smoothing* (also called Cesaro 1). Fejer smoothing produces the weighting of the coefficients of the series by

$$c(N, k) = \frac{N - k}{N}$$

Thus we have the Fejer smoothed series

$$\text{Fejer}_N(t) = \frac{a_0}{2} + \sum_{k=1}^{N} \frac{N - k}{N} [a_k \cos kt + b_k \sin kt]$$

Fejer smoothing of the rectangular pulse function is also shown in Fig. 5.2-1 and illustrates the point that the "rise time" of Fejer smoothing is very much longer than for Lanczos smoothing. Consequently, Fejer smoothing is seldom used in practice. In the limit, as N approaches infinity, this curve has nice mathematical properties; for finite values, however, it approaches its final value much too slowly.

We have examined one particular function with a jump, but it is typical of functions with jumps in them. We can expect the Gibbs phenomenon whenever a function has a discontinuity, and we can also expect the smoothing formulas to produce their corresponding effects.

5.4 COMPLEX FOURIER SERIES

By adding and subtracting Euler's identities $[i = \sqrt{(-1)}]$

$$e^{it} = \cos t + i \sin t$$
$$e^{-it} = \cos t - i \sin t$$

we get

$$\cos t = \frac{e^{it} + e^{-it}}{2}$$

$$\sin t = \frac{e^{it} - e^{-it}}{2i}$$

Since $\sin t$ and $\cos t$ are linearly independent, so are e^{it} and e^{-it}. Thus corresponding to the two functions $\sin t$ and $\cos t$, we have two complex functions e^{it} and e^{-it}; a single real frequency will give rise to two frequencies in the complex notation, one positive and one negative.

In this complex notation the Fourier series for the interval $-\pi \le t \le \pi$

$$g_N(t) = \frac{a_0}{2} + \sum_{k=1}^{N} [a_k \cos kt + b_k \sin kt]$$

becomes

$$g_N(t) = \sum_{k=-N}^{k=N} c_k e^{ikt}$$

where
$$c_k = \frac{a_k - ib_k}{2} \quad \text{for } k > 0$$

$$c_0 = \frac{a_0}{2}$$

$$c_k = \frac{a_k + ib_k}{2} \quad \text{for } k < 0$$

If $g(t)$ is an even function in t, then, by Section 4.4, $b_k = 0$ and $c_k = a_k/2$. If $g(t)$ is an odd function, $a_k = 0$ and $c_k = \pm ib_k/2$.

The complex form may be derived directly by noting that the integral

$$\int_{-\pi}^{\pi} e^{ikt} e^{-imt} \, dt = \begin{cases} 0, & k \neq m \\ 2\pi, & k = m \end{cases}$$

So if we assume the formal expansion

$$g_N(t) = \sum_{k=-N}^{k=N} c_k e^{ikt}$$

the coefficient c_m can be found by multiplying both sides by e^{-imt} (which is the complex conjugate of e^{imt}) and integrating:

$$\int_{-\pi}^{\pi} e^{-imt} g_N(t) \, dt = 2\pi c_m$$

It is easy to see (by replacing i by $-i$ if in no other way) that for a real function $g_N(t)$ we will have

$$c_k = \overline{c_{-k}}$$

where the overbar means the complex conjugate. Furthermore, the complex Fourier series is merely a notational change, a change, however, that greatly simplifies the notation. The mental adjustment necessary—thinking about complex functions with both positive and negative frequencies—is well worth the gain in the simple algebra that results.

In Section 4.5 we derived Bessel's inequality for real Fourier series expansions. For the complex expansion, we proceed in the same fashion *except* that we use a factor of $1/2\pi$ in the averaging process.

$$\frac{1}{2\pi} \int_{-\pi}^{\pi} |g(t) - \sum_{k=-N}^{k=N} c_k e^{ikt}|^2 \, dt \geq 0$$

We expand the absolute value squared as before, remembering that it is the product of the function times its conjugate. We then replace what integrals we can by the corresponding complex Fourier coefficients c_k and rearrange in order to obtain finally

$$\frac{1}{2\pi} \int_{-\pi}^{\pi} |g(t)|^2 \, dt \geq \sum_{k=-N}^{k=N} |c_k|^2$$

From Parseval's equality (if it holds) we see that the sum of the squares of the coefficients (which measures the noise propagation through a non-recursive filter; Section 1.7) can be found from the integral of the square of the transfer function.

If we return to the sampling effects discussed in Chapter 2, we still require that the complex Fourier series have two samples in the highest frequency present (if we are to avoid aliasing of some high frequencies into lower frequencies when we do the sampling). The Nyquist frequency interval now goes from $-\pi$ to π.

We have used radian measure (except in plotting) because of its convenience in the calculus operations. As noted, for most design purposes it is more convenient to measure angles in rotations. Thus we often make the change of notation

$$\omega = 2\pi f$$

The Nyquist folding frequency in rotations is now at $\frac{1}{2}$ rotation, and the basic interval of frequencies runs from $-\frac{1}{2}$ to $\frac{1}{2}$. If we are concerned with sampling rates, then the Nyquist interval is from $-\frac{1}{2}$ Hz (hertz) to $\frac{1}{2}$ Hz, measured in time units for which the sample spacing is 1. Note that in all the various notations at least two samples are always needed in the highest frequency present.

To see how the Fourier expansion is relevant to the transfer function, we convert the Fourier series from the independent variable t to the independent variable ω (or f). Given a function of ω, say $g(\omega)$, then the corresponding expansion is

$$g(\omega) = \sum_{k=-\infty}^{\infty} c_k e^{ik\omega}$$

where the c_k are given by

$$c_k = \frac{1}{2\pi} \int_{-\pi}^{\pi} g(\omega) e^{-ik\omega} \, d\omega$$

This notation matches that for the eigenvalue $\lambda(\omega)$ of Section 2.5

$$\lambda(\omega) = \sum_{k=-K}^{K} c_k e^{-i\omega k}$$

for if we set $k = -k'$, we have

$$\lambda(\omega) = \sum_{k'=-K}^{K} c_{-k'} e^{i\omega k'}$$

Beginning with Section 3.2, we used the transfer function notation

$$\lambda(\omega) = H(\omega) = \sum_{k=-K}^{K} c_{-k} e^{i\omega k}$$

The fact that the subscript on the coefficients was negative was generally concealed by the symmetry of the formulas or by avoiding labeling the co-efficients of a particular formula with their abstract symbols c_{-k}. *Thus the c_k that occur in the complex Fourier expansion of the transfer function are the same as the c_k that occur in the original definition of the transfer function.*
    For convenience, we write out the formal Fourier series in the f notation

$$g_N(f) = \sum_{k=-N}^{N} c_k e^{2\pi i k f}$$

where
$$c_k = \int_{-1/2}^{1/2} g_N(f) e^{-2\pi i k f} \, df$$

In the "real" notation we have for the variable f

$$g_N(f) = \frac{a_0}{2} + \sum_{k=-N}^{k=N} [a_k \cos 2\pi k f + b_k \sin 2\pi k f]$$

where

$$a_k = 2 \int_{-1/2}^{1/2} g(f) \cos 2\pi kf \, df$$

$$b_k = 2 \int_{-1/2}^{1/2} g(f) \sin 2\pi kf \, df$$

5.5   THE PHASE FORM OF A FOURIER SERIES

In Section 2.4 we found that although the individual coefficients a_k and b_k of a Fourier series depend for their values on the origin taken for the periodic function $f(t)$, the quantity

$$a_k^2 + b_k^2$$

is invariant under coordinate translations. For this reason, the *phase form* of a Fourier series is useful. To obtain it, we start with the usual series

$$g(t) = \frac{a_0}{2} + \sum_{k=1}^{\infty} [a_k \cos kt + b_k \sin kt]$$

and write it in the form

$$g(t) = \frac{a_0}{2} + \sum_{k=1}^{\infty} \sqrt{a_k^2 + b_k^2} \left[\frac{a_k}{\sqrt{a_k^2 + b_k^2}} \cos kt + \frac{b_k}{\sqrt{a_k^2 + b_k^2}} \sin kt \right]$$

If we define A_k and ϕ_k such that

$$\sqrt{a_k^2 + b_k^2} = A_k$$

$$\frac{a_k}{A_k} = \cos \phi_k$$

$$\frac{b_k}{A_k} = -\sin \phi_k$$

then we have

$$g(t) = \frac{a_0}{2} + \sum_{k=1}^{\infty} A_k \cos (kt + \phi_k)$$

which exhibits the phase ϕ_k and the amplitude A_k in a clear form. Each pair a_k and b_k is equivalent to the pair A_k and ϕ_k.

In the complex notation

$$g(t) = \sum_{k=-\infty}^{\infty} c_k e^{ikt}$$

we merely write the coefficients c_k in the polar form

$$c_k = A_k e^{i\phi_k}$$

and we have the corresponding phase form

$$g(t) = \sum_{k=-\infty}^{\infty} A_k e^{i(kt+\phi_k)}$$

A closely related form of the Fourier series is the *delay form*

$$f(t) = \sum_{k=-\infty}^{\infty} A_k e^{ik(t+\tau_k)}$$

where we have written

$$k\tau_k = \phi_k$$

In this notation the shift t_0 of the time origin merely adds a fixed amount t_0 to each delay term τ_k. In the phase notation the shift produces a $t_0 k$ addition to the phase ϕ_k.

5.6   GENERATING NEW FOURIER SERIES: THE CONVOLUTION THEOREMS

The basic method for finding the Fourier series of a given function $g(t)$, $-\pi < t < \pi$,

$$g(t) = \sum_{k=-\infty}^{k=\infty} c_k e^{ikt}$$

is to compute the coefficients from the integrals

$$c_k = \frac{1}{2\pi} \int_{-\pi}^{\pi} g(t) e^{-ikt}\, dt$$

This process can be both time and effort consuming for many functions.
How can we, from known (comparatively simple) expansions, derive

other expansions? Evidently if we know both

$$g(t) = \sum_{k=-\infty}^{k=\infty} c_k e^{ikt}$$

$$h(t) = \sum_{k=-\infty}^{k=\infty} d_k e^{ikt}$$

then we know

$$Ag(t) + Bh(t) = \sum_{k=-\infty}^{k=\infty} [Ac_k + Bd_k]e^{ikt}$$

How about the expansion of the product of two functions? For *real-valued* functions, we have

$$g(t)h(t) = \sum_{k=-\infty}^{k=\infty} c_k e^{ikt} \sum_{m=-\infty}^{m=\infty} d_m e^{imt}$$

Setting $n = k + m$ and rearranging terms, we have

$$g(t)h(t) = \sum_{n=-\infty}^{n=\infty} e^{int}\left[\sum_{k=-\infty}^{k=\infty} c_k d_{n-k}\right]$$

The *n*th coefficient in the expansion of the product of the two functions is, therefore,

$$\sum_{k=-\infty}^{k=\infty} c_k d_{n-k} = \sum_{k=-\infty}^{k=\infty} d_k c_{n-k}$$

This is called *the convolution* of the sequence c_k by the sequence d_k. Thus to get the coefficients for the expansion of a product of two given functions, we compute the convolutions of the coefficients.

For *complex-valued* functions, it is customary to use the product of one function times the complex conjugate of the other.

The convolution that results from multiplying two functions together suggests asking which function corresponds to multiplying the corresponding coefficients of the two expansions—that is, to which function does the series

$$\sum c_k d_k e^{ikt}$$

correspond? To answer, we introduce *the convolution of two periodic functions* $g(t)$ and $h(t)$ as

$$m(t) = \frac{1}{2\pi}\int_{-\pi}^{\pi} g(s)h(t-s)\, ds = \frac{1}{2\pi}\int_{-\pi}^{\pi} g(t-s)h(s)\, ds$$

which is symmetric in the two functions. Let us calculate the Fourier coefficients of this convolution. We have (using temporarily the notation a_k for the Fourier coefficients)

$$a_k = \frac{1}{2\pi} \int_{-\pi}^{\pi} m(t)e^{-ikt}\, dt$$

$$= \frac{1}{(2\pi)^2} \int_{-\pi}^{\pi} \int_{-\pi}^{\pi} g(s)h(t - s)e^{-ikt}e^{iks}e^{-iks}\, ds\, dt$$

$$= \frac{1}{2\pi} \int_{-\pi}^{\pi} g(s)e^{-iks}\, ds \frac{1}{2\pi} \int_{-\pi}^{\pi} h(t - s)e^{-ik(t-s)}\, dt$$

Since $h(t - s)$ is assumed to be periodic, we can shift the range of integration by shifting from the variable t to $u = t - s$, and we can see that the second integral is now d_k while the first integral, since d_k does not depend on s, is c_k. Thus we have shown that

$$a_k = c_k d_k$$

as we set out to do. In this form, we recognize that Lanczos smoothing convolves the given function $g(t)$ with a window of a continuous rectangular shape (with unit area) and produces the corresponding multiplying window in the coefficient space, the sigma factors. What we now have is the general statement of what we already knew—that due to linearity the effect of convolving a function with another function merely multiplies the coefficients of the corresponding Fourier expansions.

If in the convolution of two sequences

$$\sum_{k=-\infty}^{\infty} c_k d_{n-k}$$

we suppose that all the c_k are zero except one, say $c_0 = 1$, then the sum becomes one term—namely, d_n. As we calculate the successive values of the convolution of c_k with d_k, the terms of d_n emerge one at a time. Thus a single *impulse*—that is, a selected set of c_k having only one nonzero term—gives the *impulse response* that is simply the terms of the other sequence.

Returning to our definition of a nonrecursive filter

$$y_n = \sum_{k=-\infty}^{\infty} c_k x_{n-k}$$

if we use the impulse function $x_k = 0$ except for $x_0 = 1$, then we will obtain values for the y_n that are simply the coefficients of the filter, c_n, in order; the impulse produces a response that gives the coefficients of the filter.

Conversely, if we know the impulse response of a nonrecursive filter, then we know the coefficients and hence the filter. Thus the impulse response defines the filter and plays a fundamental role in the theory.

5.7 MORE ON THE GIBBS PHENOMENON

Using the second form of the convolution theorem, we can now see the Gibbs phenomenon in a different light. Given a function represented as a Fourier series

$$g(t) = \sum_{k=-\infty}^{\infty} c_k e^{ikt}$$

the act of truncating this series to

$$g_N(t) = \sum_{k=-N}^{k=N} c_k e^{ikt}$$

is the same as multiplying the coefficients c_k by the numbers $(2N + 1$ values each equal to 1)

$$0, 0, 0, 1, 1, \ldots, 1, 0, 0, 0, \ldots$$

or

$$d_k = \begin{cases} 1, & |k| \leq N \\ 0, & |k| > N \end{cases}$$

What function $h(t)$ has these $2N + 1$ nonzero coefficients d_k? Clearly,

$$h(t) = \sum_{k=-N}^{k=N} e^{ikt} = e^{-iNt} + e^{-i(N-1)t} + \cdots + e^{iNt}$$

In Section 3.2 we summed this geometric progression

$$h(t) = \frac{e^{i(N+1/2)t} - e^{-i(N+1/2)t}}{e^{it/2} - e^{-it/2}}$$

$$= \frac{\sin (N + 1/2)t}{\sin (t/2)} \qquad (|t| < \pi)$$

For large N, this is a rapidly oscillating function with the maximum value $(2N + 1)$ at $t = 0$ and one that falls off rapidly in amplitude as the denominator grows larger. See Fig. 5.7-1 for the case $N = 5$, (unmodified rectangular window).

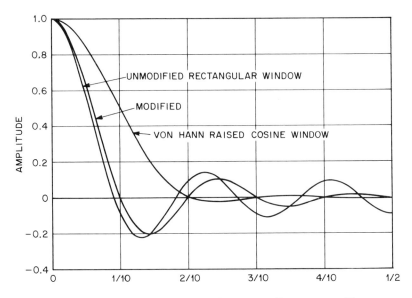

FIGURE 5.7-1   COMPARISON OF THE FREQUENCY RESPONSE OF THREE
WINDOW FUNCTIONS—5 TERMS

Thus truncating a Fourier series $g(t)$ is equivalent (by the second convolution theorem) to convolving the given function $g(t)$ with the function

$$h(t) = \frac{\sin (N + 1/2)t}{\sin t/2}$$

As shown in Fig. 5.7-2, in the particular case of a square wave, $g(t)$, as the
wiggles of $h(t)$ hit the rise of the square wave, the convolution (which is the
integral of the product of the functions) will produce exactly the wiggles of
the Gibbs phenomenon. Thus we see the cause of the Gibbs phenomenon
in a new way.

5.8   MODIFIED FOURIER SERIES

Instead of truncating a Fourier series by multiplying its coefficients by

$$\ldots 0, 0, 0, 1, 1, 1, \ldots, 1, 1, 0, 0, 0, \ldots$$

we can use the sequence having $\frac{1}{2}$ as the two end values

$$\ldots 0, 0, 0, \tfrac{1}{2}, 1, 1, \ldots, 1, \tfrac{1}{2}, 0, 0, 0, \ldots$$

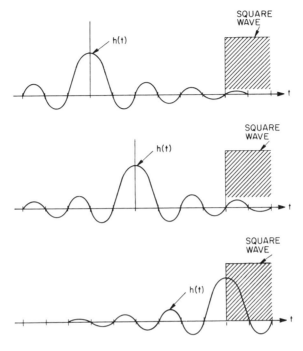

FIGURE 5.7-2 GIBBS PHENOMENON

For this *modified rectangular window*, we get not the earlier $h(t)$ but one that is decreased by

$$\frac{1}{2}(e^{iNt} + e^{-iNt}) = \cos Nt$$

So our new $h(t)$ is

$$h(t) = \frac{\sin (N + 1/2)t}{\sin t/2} - \cos Nt$$

Expanding the first term, we get

$$h(t) = \frac{\sin Nt \cos t/2 + \cos Nt \sin t/2 - \cos Nt \sin t/2}{\sin t/2}$$

or

$$h(t) = \frac{\sin Nt \cos t/2}{\sin t/2}$$

Comparing this (Fig. 5.7-1) with the earlier unmodified rectangular window, we see a slight change in the high-frequency term but also an extra factor of $\cos t/2$. This extra factor begins with the value of 1 at $t = 0$ and decreases to 0 at $t = \pi$ (the folding frequency). Consequently, this convolving window is significantly smaller at the ends and thus produces "slightly better results" in many situations. However, working out the corresponding over-shoot for $N \rightarrow \infty$, it turns out to be the same; the kernels are almost the same at the center but differ significantly near the ends.

As a result of the theory of convergence of a Fourier series at a point of discontinuity, it is not surprising that this modified rectangular window can give better results. But it does raise a central point, one that will be discussed in detail in the future—how to deal with a finite piece of an infinite record. The use of a rectangular window clearly has bad properties, whereas even the simplest modification produces some improvement.

5.9 THE VON HANN WINDOW: THE RAISED COSINE WINDOW

When truncating a Fourier series, it is exactly the same as looking at the original function through a window. Ideally, we would like a tall, narrow window of unit area so that what we "see" is very close to the original function at the middle of the (symmetric) window. Unfortunately, we also want to use as few terms as possible in the Fourier series, which means using a wide window.

The width of a window may be measured by the distance between the closest zeros on each side of the center lobe—the main lobe of the window. But it can also be measured in a different sense by the *size* of the lobes that are farther away from the center. It will later be seen that the main lobe gives rise to the transition band (a region between a stop and a pass band), and the side lobes give rise to the wiggles (the Gibbs phenomenon), which may be regarded as "contamination" from parts of the function that are adjacent to where we are looking.

Because we are interested in transfer functions (although the results apply to arbitrary functions), we shift to the f notation.

The simple rectangular window (Section 5.6 with $t = 2\pi f$) gives rise to the convolving window through which we see the function

$$h(f) = \frac{\sin\,[\pi(2N + 1)f]}{\sin\,\pi f}$$

The modified rectangular window (Section 5.7) merely takes one-half of the end values, but it moves the first zeros out from

$$f = \pm\frac{1}{2N + 1} \quad \text{to} \quad f = \pm\frac{1}{2N}$$

It also puts in an extra cosine factor that steadily decreases the sizes of the side lobes that are farther and farther away from the center of the window. For $N = 5$ (11 terms in the final filter), Fig. 5.7-1 illustrates this change; the slight loss in narrowness is greatly compensated for by the decrease in the heights of the other lobes. The sigma factors from the Lanczos window also show how tapering the weights applied to the coefficients of the Fourier series can greatly decrease the heights of the side lobes. The third curve in the figure will be examined next.

The foregoing remarks suggest examining a more severe weighting of the Fourier coefficients that we keep in the truncation process. This *von Hann window* is also called the "raised cosine window" from its definition

$$w_k = \begin{cases} \dfrac{1 + \cos \pi k/N}{2}, & |k| \leq N \\[2mm] 0, & |k| \geq N \end{cases}$$

The weighting sequence of $2N + 1$ terms, when viewed as a continuous function, not only vanishes at the ends but is also tangent there.

The corresponding transform in the frequency domain of the von Hann window is, by definition, the function

$$h(f) = \sum_{k=-N}^{k=N} w_k e^{2\pi i k f}$$

$$= \frac{1}{4} \sum_{k=-N}^{k=N} (e^{i\pi k/N} + 2 + e^{-i\pi k/N}) e^{2\pi i k f}$$

Since $w_N = 0$, taking half the end values has no effect, and the results of Section 5.7 can be applied to the three separate terms in the summation. This step gives

$$h(f) = \frac{1}{4} \left\{ \frac{\sin [\pi(f + 1/2N)2N] \cos [\pi(f + 1/2N)]}{\sin [\pi(f + 1/2N)]} + 2 \frac{\sin \pi f 2N \cos \pi f}{\sin \pi f} \right.$$
$$\left. + \frac{\sin [\pi(f - 1/2N)2N] \cos [\pi(f - 1/2N)]}{\sin \pi(f - 1/2N)} \right\}$$

which may be written as

$$h(f) = \frac{1}{4} \left\{ \frac{\sin \pi(2Nf + 1) \cos [\pi(f + 1/2N)]}{\sin [\pi(f + 1/2N)]} + \frac{2 \sin \pi f 2N \cos \pi f}{\sin \pi f} \right.$$
$$\left. + \frac{\sin \pi(2Nf - 1) \cos [\pi(f - 1/2N)]}{\sin [\pi(f - 1/2N)]} \right\}$$

Expanding the sines in the numerators of the end terms, we get ($\sin \pi = 0$ and $\cos \pi = -1$)

$$h(f) = \frac{\sin 2\pi Nf}{4}\left\{-\cot \pi\left(f + \frac{1}{2N}\right) + 2\cot \pi f - \cot \pi\left(f - \frac{1}{2N}\right)\right\}$$

Next, applying the trigonometric identity

$$\cot (a + b) + \cot (a - b) = \frac{2 \sin a \cos a}{\sin^2 a - \sin^2 b}$$

gives

$$h(f) = \frac{\sin 2\pi Nf}{4}\left\{\frac{2 \cos \pi f}{\sin \pi f} - \frac{2 \sin \pi f \cos \pi f}{\sin^2 \pi f - \sin^2 (\pi/2N)}\right\}$$

This result may be rewritten in a form that shows the dependence on the parameters as follows:

$$h(f) = \frac{\sin 2\pi Nf \cos \pi f}{2 \sin \pi f}\left[\frac{1}{1 - \left(\dfrac{\sin \pi f}{\sin \pi/2N}\right)^2}\right]$$

The zeros of the denominator in the braces ($f = \pm 1/2N$) are compensated by corresponding zeros in the front multiplier, and $h(f)$ has the finite value $N/2$ at $f = \pm 1/(2N)$. As $f \rightarrow 0$,

$$h(0) \longrightarrow \frac{2\pi N}{2\pi} = N$$

The first zeros of $h(f)$ occur at $f = 1/N$. Thus the window is twice as wide as the modified window, *but* the side lobes are greatly reduced, as shown in Fig. 5.7-1. In a sense, this window is the sum of three modified windows, one at $f = -1/2N$ of size $1/4$, one at $f = 0$ of size $1/2$, and, finally, one at $f = 1/2N$ of size $1/4$.

5.10 HAMMING WINDOW: RAISED COSINE WITH A PLATFORM

From Fig. 5.7-1 it is evident that the von Hann window and the modified window have opposite signs in the side lobes. This situation immediately suggests that a small amount of the modified window added to the van Hann window could be used to reduce the maximum that occurs in the side lobes.

The outcome is the *Hamming window* (sometimes written as "hamming window" [BT, p. 98]), which has coefficients 0.23, 0.54, 0.23 instead of the von Hann coefficients 0.25, 0.50, 0.25.

To find the coefficients of the Hamming window, we merely take a weighted sum of the modified rectangular window and the von Hann window and use an optimizing routine (see Section 9.9) to find the weights that minimize the maximum of the side lobes (tails) of the window. This is the Chebyshev criterion, which will be discussed in Chapter 12. The value obtained for the weights depends, of course, on the value of N used in the weighting window $\{w_k\}$. As can be seen from Fig. 5.10-1, the values of a and b in the Fourier expansion vary slowly with $1/N$ for reasonably long runs of data. Notice that $2a + b = 1$ for all values of $1/N$.

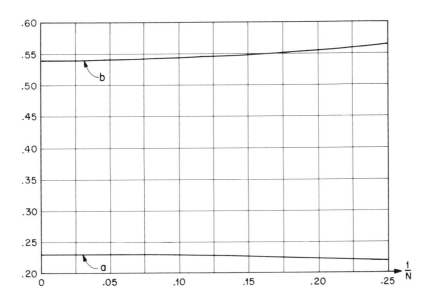

FIGURE 5.10-1   COEFFICIENTS FOR THE HAMMING WINDOW $2\,a\cos(\pi n/N) + b$

Because of the central role of least squares, the idea of minimizing the integral of the square of the side lobes *relative to* the integral of the square of the main lobe will probably occur to most people. The result of this optimization appears in Fig. 5.10-2, which gives the small corrections to the Hamming window in terms of a quantity labeled d. The reason for the peculiar shape is revealed by a detailed look at the tails of the window. Both the Chebyshev minimum maximum error and the least-squares error are given in Fig. 5.10-3, where we see the large negative first lobe of the least-squares optimization.

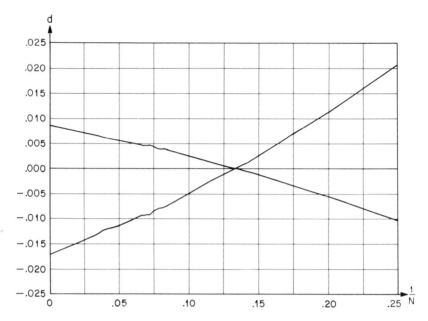

FIGURE 5.10-2   Coefficients for the Optimal Least Squares Window,
$a = 0.23 + d, b = 0.54 - 2d$

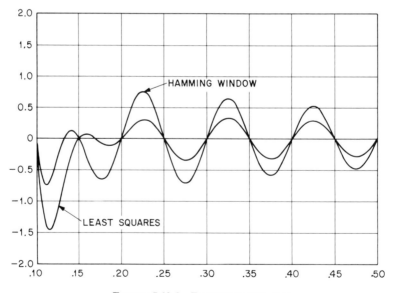

FIGURE 5.10-3   Frequency Response

The general shape of the weighting factors of the von Hann and Hamming windows is illustrated in Fig. 5.10-4. The figure shows why the Hamming window is often referred to "as a raised cosine on a platform."

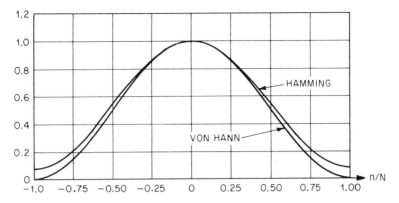

FIGURE 5.10-4   WEIGHT FACTORS FOR HAMMING AND VON HANN WINDOWS

5.11   REVIEW OF WINDOWS

Because windows are the most confusing part of digital filter theory, let us review the situation.

We began with a continuous signal $x(t)$ and sampled it at unit intervals in order to obtain the set of measurements $\{x_n\}$. When we limited the signal to the range $-N \leq n \leq N$, it follows from the second convolution theorem (Section 5.6) that this process was equivalent to "smearing the spectrum of the signal" by effectively looking at the true function $x(t)$ through the convolving window (Section 5.7)

$$h(t) = \frac{\sin (N + 1/2)t}{\sin t/2}$$

If instead of merely looking at a section of the sequence of data points $\{x_n\}$, we also weigh it with weights $\{w_n\}$—that is, we use the sequence $\{w_n x_n\}$ in place of $\{x_n\}$— then we get a different window. Merely taking half the end values changes the convolving window to

$$h(t) = \frac{\sin Nt \cos t/2}{\sin t/2}$$

Although doing so somewhat decreases the side lobes, they are still large enough to produce significant distortions of the original transform.

Further modification of the weights leads to the von Hann window (Section 5.9) with significantly smaller side lobes but also twice the width in the main lobe (which means a loss in distinguishing close details). A slight additional modification leads to the Hamming window (Section 5.10), which has the smallest extreme value in the side lobes.

However, we can also use the first convolution theorem. Thus if we use one of the sequences

$$\{\tfrac{1}{4}, \tfrac{1}{2}, \tfrac{1}{4}\}$$
$$\{0.23, 0.54, 0.23\}$$

as a smoothing formula on the original data $\{x_n\}$, we effectively multiply the transform by the corresponding continuous windows (Fig. 5.10-4). These windows tend to remove the higher frequencies.

Various types of windows may be used in either of the two ways, on the function $x(t)$ or on the transfer function $H(\omega)$, and an elementary textbook is not the place to discuss all of them (see IEEE-1).

APPENDIX 5.A

A sequence S_N is said to converge to S ($S_N \longrightarrow S$) if no matter how close one asks (given an $\epsilon > 0$), it is possible to find a place N_0 (there exists an N_0) such that for all greater values of N (for all $N \geq N_0$) the values of S_N are all closer than required ($|S_N - S| < \epsilon$).

The convergence of a series is reduced to the convergence of a sequence by the simple device of considering the sequence of partial sums of the series.

If we now think of a series whose terms depend on a variable, say t, then for each t we can consider the convergence. In this situation, the N_0 will depend on *both* ϵ and t,

$$N_0 = N_0(\epsilon, t)$$

In a given interval (open or closed), it may be possible to find an N_0 *independent* of t that will meet the convergence condition for a given ϵ. If so, the series is said to be *uniformly convergent*; otherwise the series is not uniformly convergent.

6

Design of
Nonrecursive Filters

6.1 INTRODUCTION

Now that we have the necessary mathematical theory, we are ready to design digital filters. Recall that the typical smoothing filter was a *lowpass filter*, meaning that the low frequencies "pass" through and the high frequencies are "stopped" (eliminated), with a transition zone between the pass and stop bands (or frequencies). (See Fig. 6.1-1.) In Chapter 3 we designed such filters by choosing a finite, symmetric set of coefficients. We selected a digital filter of the form

$$y_n = \sum_{k=-N}^{k=N} c_k x_{n-k} \qquad (c_k = c_{-k})$$

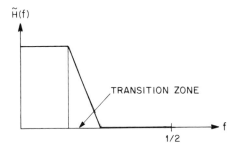

FIGURE 6.1-1 LOWPASS FILTER

Had we wanted to interpolate missing data, we would have picked much the same type of digital filter but set $c_0 = 0$.

A *highpass filter* is, of course, the opposite of a lowpass filter: it passes the high frequencies and stops the low (Fig. 6.1-2). It is simply the difference between an "all-pass filter" $y_n = x_n$ and a lowpass filter.

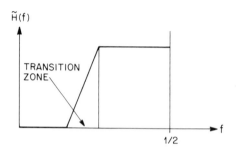

FIGURE 6.1-2   HIGHPASS FILTER

There are also *bandpass* and *bandstop* filters. A bandpass filter is often used to study a part of a spectrum. Frequently a very narrow bandstop filter is called a *notch filter*. Among other uses, a notch filter (Fig. 6.1-3) is used to remove the ever present 60 Hz that comes from our electrical power distribution system (in the United States; in England it is 50 Hz).

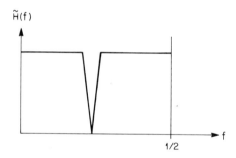

Figure 6.1-3   NOTCH FILTER

Differentiation filters require odd symmetry in the coefficients (and thus lead to only sine terms in the Fourier expansion); they have $c_k = -c_{-k}$ and $c_0 = 0$. Integration cannot be done by nonrecursive filters.

In Section 4.4 we showed that any function can be written as the sum of an even and an odd function. Similarly, the following identity

$$c_k = \frac{c_k + c_{-k}}{2} + \frac{c_k - c_{-k}}{2}$$

shows that any digital filter can be written as the sum of a smoothing (even) filter and a differentiating (odd) filter. A smoothing filter can be viewed as a linear combination of the sums of symmetrically placed data, while a differentiating filter uses differences. Clearly, they are simply the cosine and sine terms of the general Fourier series expansion.

Given a symmetric digital filter, $c_k = c_{-k}$, we substitute $x_n = e^{i\omega n}$ into the equation and obtain the transfer function $H(\omega)$. As a result of the symmetry of the coefficients, we have a Fourier series in cosines and a Nyquist interval of $-\pi \le \omega \le \pi$. In the frequency notation f, the Nyquist interval is $-\frac{1}{2} \le f \le \frac{1}{2}$. The transfer function $\tilde{H}(f)$ is symmetric about $f = 0$. A graph of the values of this Fourier series gives the curve of the transfer function.

We can reverse these steps (Fig. 6.1-4). Given an arbitrarily shaped (sym-

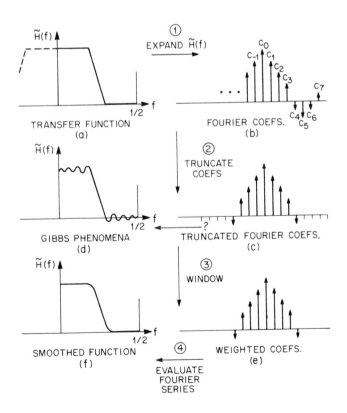

FIGURE 6.1-4

metric) transfer function [part (a) of Fig. 6.1-4], we can find, using the methods of Sections 4.4 and 5.4, the coefficients of the corresponding cosine Fourier series [part (b)]. This Fourier series will, in general, have an infinite number of coefficients; in practice, we want a finite filter, and so we are forced to truncate the coefficients past some subscript value N [part (c)]. But the truncation produces the Gibbs phenomenon, as discussed in Sections 5.2, 5.6, and 5.7 [part (d)]. To avoid it, we apply the Lanczos window (Section 5.3), which multiplies the coefficients by the sigma factors [part (e)]. This process, in turn, produces the smoothed transfer function [part (f)]. The total effect on the coefficients c_k of the complex Fourier series expansion of the transfer function is that we multiply the c_k by the sigma factors $\sigma(N, k)$ for $|k| \leq N$ and by 0 for $|k| > N$. The sigma factors are, therefore, the coefficient multiplying window. In the transform, this window is the combination of first creating the Gibbs phenomenon and then doing the Lanczos smoothing.

The use of the von Hann or Hamming windows greatly reduces the ripples in the final transfer function but also doubles the width of the transition band.

In our basic design method (illustrated in Fig. 6.1-4) the step going to part (b) finds the Fourier coefficients that are the least-squares fit (Section 4.4). The sigma factors alter this basic least-squares fit. In the next section the method will be illustrated by a specific example.

To summarize our first method of design, given a symmetric transfer function $H(\omega)$, we perform the following steps: find the first N Fourier cosine coefficients $a_0, a_1, \ldots, a_{N-1}$; then multiply these coefficients by the corresponding sigma factors, using the same value N; convert the resulting coefficients to the appropriate c_k (of the filter) by dividing by 2 (beware of the constant term); and, finally, plot the resulting transfer function to check the results.

6.2   A LOWPASS FILTER DESIGN

We will now design a lowpass filter giving some numerical details. Our basic design method is that of Fig. 6.1-4. As a first, concrete example, we select the transfer function [Fig. 6.2-1 curve (a)]

$$\tilde{H}(f) = \begin{cases} 1, & 0 < |f| < 0.2 \\ 0, & 0.2 < |f| < 0.5 \end{cases}$$

with, of course

$$\tilde{H}(-f) = \tilde{H}(f)$$

By direct computation of the integrals for the coefficients of the Fourier series (Sections 4.4 and 5.4), all the $b_k = 0$, and

$$a_k = \frac{2}{\pi} \int_0^\pi H(\omega) \cos k\omega \, d\omega$$
$$= 4 \int_0^{1/2} \tilde{H}(f) \cos 2\pi k f \, df$$

Thus we have for the coefficients of our particular ideal transfer function of a lowpass filter

$$a_k = 4 \int_0^{0.2} \cos 2\pi k f \, df = \frac{2}{\pi k} \sin 0.4\pi k$$

and the corresponding Fourier series [Fig. 6.1-4 (b)] is

$$\tilde{H}(f) = \frac{4}{10} + 2 \sum_{k=1}^{\infty} \left[\frac{\sin (4k\pi/10)}{\pi k} \right] \cos 2\pi k f$$

For practical reasons, we *truncate the infinite series to a finite length*. We will use $N = 5$ terms, and so we set $N - 1 = 4$ in the summation

$$\tilde{H}(f) = \frac{4}{10} + 2 \sum_{k=1}^{k=4} \left[\frac{\sin (4k\pi/10)}{\pi k} \right] \cos 2\pi k f$$

We are now at part (c) in Fig. 6.1-4 and face the Gibbs phenomenon at part (d). To reduce the size of the ripples, [Fig. 6.2-1 curve (d)] we use the Lanczos window, which means multiplying the coefficients of the Fourier expansion by the appropriate sigma factors (Section 5.2)—in this case,

$$\sigma(5, k) = \frac{\sin (\pi k/5)}{\pi k/5}$$

Notice that $\sigma(5, 5) = 0$ and that we would have removed the term a_5 had we tried to keep it. Thus for the modified transfer function, we have

$$\tilde{H}(f) = \frac{4}{10} + 2 \sum_{k=1}^{k=4} \left(\frac{\sin (\pi k/5)}{\pi k/5} \right) \left[\frac{\sin (4k\pi/10)}{\pi k} \right] \cos 2\pi k f$$

Having used the Lanczos window, we now have the transfer function at part (e) and the Fourier coefficients at part (f) in Fig. 6.1-4. Thus we have our

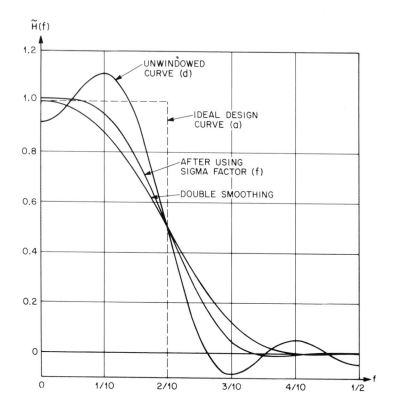

FIGURE 6.2-1 LOWPASS FILTER, $N = 5$

filter design as shown in Fig. 6.2-1 curve (f) for $N = 5$. See Fig. 6.2-2 for the case of $N = 10$. The third curve in each figure will be discussed in the next section. The digital filter coefficients c_k are half those in the cosine expansion except for the constant term.

Exercises

6.2-1   Design a lowpass filter using $N = 5$, $\tilde{H}(f) = 1$ for $|f| < \frac{1}{5}$, and 0 elsewhere.

6.2-2   Design a highpass filter with $N = 4$ that passes the upper half of the Nyquist interval.

6.2-3   Design a bandpass filter that passes the middle third of the Nyquist interval (use $N = 6$).

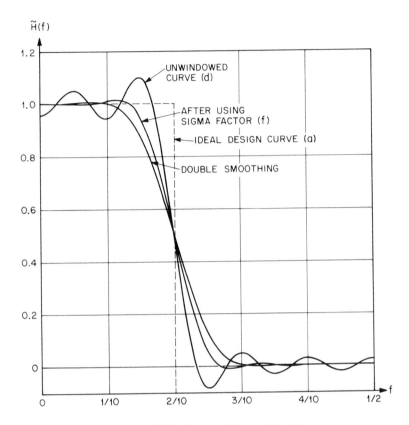

FIGURE 6.2-2   LOWPASS FILTER, $N = 10$

6.2-4   Design a highpass filter with $\tilde{H}(f) = 1$, $|f| \geq \frac{4}{5}$ and 0 elsewhere (use $N = 5$).

6.2-5   Design a bandpass filter with $\tilde{H}(f) = 1$ for $\frac{2}{5} \leq |f| \leq \frac{3}{5}$ and 0 elsewhere (use $N = 5$).

6.3   CONTINUOUS DESIGN METHODS: A REVIEW

Having designed one specific filter, let us turn, using our earlier developments, to a slightly more general approach to a lowpass filter design.

First, we start with an arbitrary width for the pass band; the pass band will be from zero to f_s and the stop band will go from f_s to $\frac{1}{2}$ (measured in f, rotations). Direct calculation of the Fourier coefficients (remember that this transfer function is an even function of f) gives the expansion

$$\tilde{H}(f) = 2f_s + \sum_{k=1}^{k=\infty} \left[\frac{2}{\pi k} \sin 2\pi k f_s \right] \cos 2\pi k f$$

This expansion agrees with the result of Section 6.2 when $f_s = 0.2$.

After truncating it to a finite number of terms (applying a rectangular multiplying window to the coefficients), we have the Gibbs phenomenon in the frequency domain. A rectangular convolution window, the Lanczos window, is used to smooth this effect. Thus we apply the sigma factors. In Section 6.2 we chose the number of terms $N = 5$ (and $N = 10$) and adjusted the width of the window to the width of the ripples in the Gibbs phenomenon.

Consider, now, selecting the transition width and using it to determine the number of terms. We choose a general rectangular Lanczos convolving window of width Δ and unit area. Earlier we showed that any linear operation on the function will appear as a multiplication of the Fourier coefficients by some constants, just as it happened in the Lanczos window, or in the convolution of two functions. By carrying out the algebra and trigonometry in detail, we will find that, corresponding to the sigma factors, we get the factors

$$\frac{\sin \pi k \, \Delta}{\pi k \, \Delta}$$

as multipliers of the corresponding coefficients of the Fourier series. We naturally truncate where these factors first become small.

It is important to note that we could also view this whole process as first smoothing the transfer function. When in our mind's eye we picture the convolution of the rectangular window (dot-dash lines in Fig. 6.3-1) going across

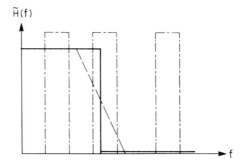

FIGURE 6.3-1 A RECTANGULAR WINDOW ON A DISCONTINUITY

the original transfer function, we see that (assuming that the window is narrow enough) in the beginning we obtain the same constant value *until* the window begins to touch the discontinuity (dashed line). As the window crosses the discontinuity, the smoothed value decreases linearly. When the window is completely across the discontinuity, the smoothed value is zero. Thus we may, as an alternate to the original problem of finding an approximation to the discontinuous function, approximate the function with a straight-line interpolation between the two parts, the width of the transition zone being exactly the width of the rectangular window used. (See Fig. 6.3-1.) In this way, we see that whether we take the original function and, after truncation, smooth with the sigma factors or imagine the rectangular window being convolved with the original function and then fit the resulting smoothed function, we must get the same result (we, of course, truncate the Fourier series in both cases).

The results of the two approaches appear slightly different, although the effect is the same. Either we truncate and smooth or we smooth the curve to get a more rapidly converging series and then truncate. The width of the window determines the width of the transition band between the pass and stop bands.

If using the window once produces some smoothness in the transfer function, applying it twice should produce a transfer function that has a continuous first derivative and hence more convergence in the resulting Fourier series. Applying the rectangular window twice causes the sigma factors to occur twice; that is, we get the square of the sigma factors. Of course, the double application of the window results in a wider transition band. (See the third curves in Figs. 6.2-1 and 6.2-2.) Again we can look at the process several ways. We may, for instance, first ask about the result of convolving the window with itself and then applying the result once. A few moment's thought shows that the convolution of the rectangular window with itself will result in a triangular-shaped window whose base is twice the original window width and whose apex is in the middle (Fig. 6.3-2). Thus a triangular window is equivalent to double application of the rectangular window.

The "carpentry" possibilities of windows are endless. We can start with other than rectangular-shaped windows. Such windows are called *tinted*; a variable amount of the original function gets through at various points. One example is our triangular window above. In short, we can use any reasonable function that we please, supposing, for convenience, that it has unit area under the curve, and any combination, via convolutions, can be tried. The fact that we will always obtain numerical factors that multiply each frequency is derived from our earlier observation that any linear operation on a fre-

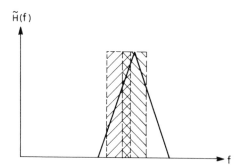

$\tilde{H}(f)$

FIGURE 6.3-2 A TRIANGULAR WINDOW FROM TWO RECTANGULAR WINDOWS

quency gives the same frequency but changes the coefficients of the Fourier expansion by a corresponding multiplicative factor (depending on the frequency). For symmetric windows, we can expect that operating on a cosine expansion will produce only cosine terms; but if there are asymmetries in the window, then both the sine and cosine terms must be viewed together.

Exercises

6.3-1 Smooth the rectangular pulse function with the Lanczos window and then compute the Fourier coefficients directly.

6.3-2 Smooth the rectangular pulse function with the triangular window and then compute the Fourier coefficients.

6.3-3 Carry out the derivation of the window of width Δ.

6.4 A DIFFERENTIATION FILTER

As another illustration of this general method for filter design, let us examine how this general method of fitting a given transfer function by a Fourier series handles the problem of designing a filter to estimate the derivative of some data. It is immediately clear from the expression for the derivative

$$\frac{d}{dt}[e^{i\omega t}] = i\omega e^{i\omega t}$$

that we want to approximate the function

$$H(\omega) = i\omega$$

If we pick the coefficients of the filter to have odd symmetry, that is

$$c_{-k} = -c_k \qquad \text{(for all } k\text{)}$$

then we see that since

$$c_k(e^{ik\omega} - e^{-ik\omega}) = 2ic_k \sin k\omega$$

we will have a sine series and that it will be purely imaginary as required. Thus the digital filter

$$y_n = \sum_{k=-N}^{N} c_k x_{n-k}$$

with $c_{-k} = -c_k$ leads to the sine series

$$H(\omega) = [2c_1 \sin \omega + 2c_2 \sin 2\omega + \cdots + 2cN \sin N\omega]i$$

In examining such a filter, we see that it is, in fact, some linear combination of differences of symmetrically placed values of the function (estimates of the derivative)

$$c_k(x_{n+1} - x_{n-k})$$

which is what we would expect once we think about it. Furthermore, it is clear that the process of differentiation amplifies high frequencies much more than low frequencies. Since high frequency is often the noise, this means that the filter that we design should probably cut off the frequencies past some value ω_c. Thus we will make the Fourier sine series approximate the function

$$H(\omega) = \begin{cases} i\omega, & |\omega| < \omega_c \\ 0, & |\omega| > \omega_c \end{cases}$$

We compute the coefficients by the usual formulas

$$b_k = \frac{1}{\pi} \int_{-\pi}^{\pi} H(\omega) \sin k\omega \, d\omega = \frac{2}{\pi} \int_{0}^{\omega_c} i\omega \sin k\omega \, d\omega$$

$$b_k = \frac{2i}{\pi} \left(\frac{\sin k\omega_c}{k^2} - \frac{\omega_c \cos k\omega_c}{k} \right)$$

As a check on the algebra, put $\omega_c = \pi$. We get

$$b_k = \frac{-2i \cos \pi k}{k} = \frac{2i}{k}(-1)^{k+1}$$

which agrees with the result of Section 4.3. Thus we have the infinitely long filter that we must truncate, and so we face the Gibbs effect. For the moment we simply use the rectangular Lanczos window, and doing so produces the corresponding sigma factors in the expansion. To illustrate, we choose $f_c = \frac{2}{10}$. For values of $N = 5, 10$, we have the corresponding curves plotted in Figs. 6.4-1 and 6.4-2. The straight line is the ideal, the most wiggly curve is the truncated Fourier series, which has large errors near $f = 0$, and the lower curve is the final, smoothed filter. For additional details on differentiating filters, see [K, pp. 218–285].

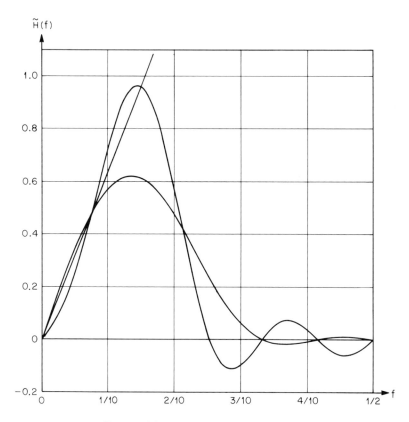

FIGURE 6.4-1   DIFFERENTIATOR WITH $N = 5$

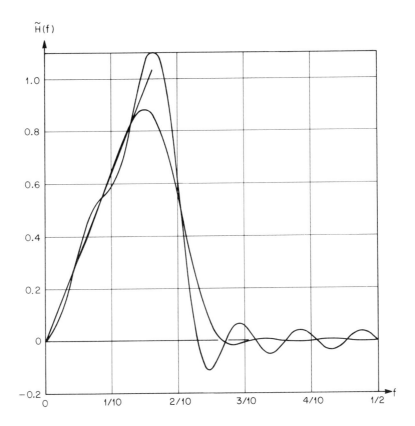

FIGURE 6.4-2   DIFFERENTIATOR WITH $N = 10$

Exercises

6.4-1   In [L, p. 321] a "low-noise" differentiation filter is proposed that (a) has unit slope at $f = 0$ and (b) minimizes the sum of the squares of the coefficients. Show that this filter is

$$3 \sum_{k=-N}^{k=N} k \frac{x_{n-k}}{N(N + 1)(2N + 1)}$$

(*Hint:* Use the Lagrange multiplier method.) See Fig. 6.4-3.

6.4-2   Super Lanczos low-noise differentiators. In Exercise 6.4-1, if we require that the tangency at $f = 0$ be of third order, show that the first few

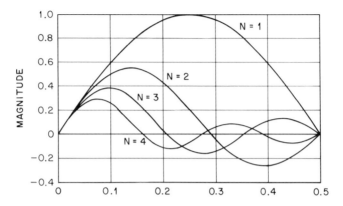

FIGURE 6.4-3   LANCZOS DIFFERENTIATING FILTERS

filters have transfer functions

$$\frac{8 \sin \omega - \sin 2\omega}{6}$$

$$\frac{58 \sin \omega + 67 \sin 2\omega - 22 \sin 3\omega}{126}$$

$$\frac{126 \sin \omega + 193 \sin 2\omega + 142 \sin 3\omega - 86 \sin 4\omega}{594}$$

See Fig. 6.4-4.

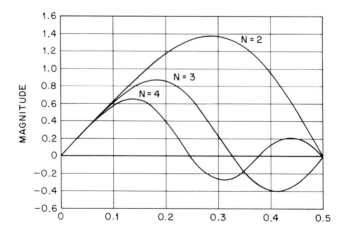

FIGURE 6.4-4   SUPER LANCZOS DIFFERENTIATING FILTERS

6.5   TESTING THE DIFFERENTIATING FILTER
ON DATA

As a test of the differentiating filter, let us make up some artificial data. We first take the straight line

$$x_n = \frac{n}{50} \qquad (n = 0, 1, \ldots, 50)$$

whose derivative is, of course, $\frac{1}{50}$. To calculate the response of the filter to this input, we need only note that since $c_k = -c_{-k}$ and $c_0 = 0$, the sum of the coefficients is zero. Therefore we can add any constant that we like to the data without affecting the answer. So we may take the straight line $x_k = k$ and test the filter at the origin of the data because it is the same as at any other place. We will then have for the output

$$\sum_{k=-N}^{k=N} c_k x_k = 2 \sum_{k=1}^{k=N} k c_k$$

Without the sigma factors, we get for the value of this sum, using $N = 5$, the number $0.5165/50$ and, for $N = 10$, the number $0.5568/50$ (where we have kept the factor 50 so that the slope of the line is taken care of and the ideal value of the numerator is one). Our choice of $N = 5, 10$ is guided by the way that we can reasonably expect the harmonics of the Fourier series to interact with the edge of the cutoff frequency at $\frac{2}{10}$.

The failure of the filter to give the value $\frac{1}{50}$ is directly traceable to the fact that at $f = 0$ the slope of the transfer function is not even near one (Figs. 6.4-1 and 6.4-2) (highest curve), for if it were, the calculated sums would be equal to one.

If we use the Lanczos sigma factors (lower curve), we obtain much better results, as the figures show. The initial slope is $1.047/50$ for $N = 5$ and $1.020/50$ for $N = 10$. The straight lines show the ideal slope.

Remember, however, that we designed the filter to cope with noise. Consequently, we consider further experiments. We add random noise in two forms; first,

$$\frac{N}{50} \pm \frac{1}{100} \begin{cases} + & \text{if random number} > \frac{1}{2} \\ - & \text{else} \end{cases}$$

and a second noise

$$\frac{N}{50} + \frac{1}{100} \text{ (random number with range } -1 < x < 1)$$

Figures 6.5-1 to 6.5-4 show the results of these experiments, where we have plotted (a) the original input data sloping upward, (b) ten times the derivative so that the fluctuations can be seen easily, and (c) when using the sigma factors, we also added $\frac{1}{2}$ to shift the display so that it is easily seen.

The set of experiments shows that when a filter is asked to do several things, such as both differentiate and reject a significant amount of noise, it compromises and does neither too well. The experiments also show that even simple filters can be made to do complex jobs. The behavior on a straight line was not one of the design criteria; rather it was designed to work on pure frequencies and sums of frequencies, and on them it does as expected, for we can see from the transfer function exactly what happens to the amplitude of each frequency.

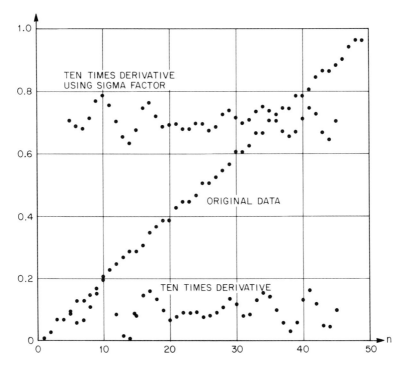

FIGURE 6.5-1   NOISE 1, $N = 5$

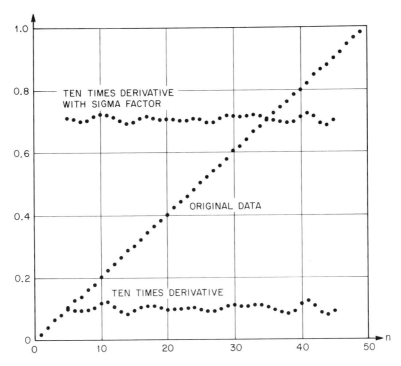

FIGURE 6.5-2 Noise 2, $N = 5$

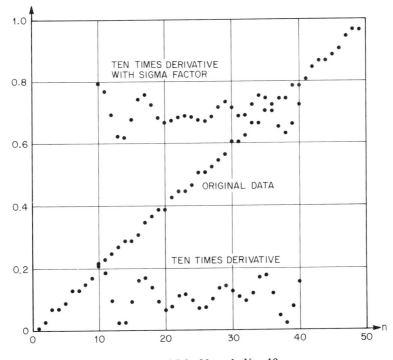

FIGURE 6.5-3 Noise 1, $N = 10$

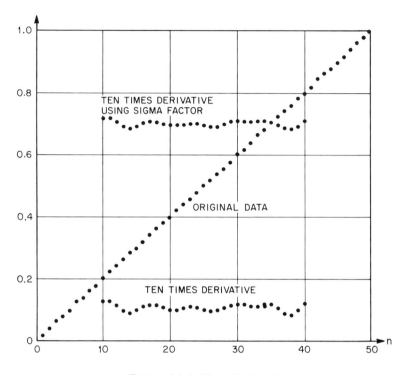

FIGURE 6.5-4 NOISE 2, $N = 10$

6.6 NEW FILTERS FROM OLD ONES: SHARPENING A FILTER

It soon occurs to most students in filter theory that if using a lowpass (or highpass or even a bandpass) filter is good, then perhaps running the data through the filter twice might be even better. As we shall see, this process will

1. approximately double the errors in the pass band.
2. square the error in the stop band.
3. leave the transition bands the same.
4. approximately double the length of the equivalent filter (and hence the loss of data at each end).

Of course, the equivalent single filter can be obtained by convolving the sequence of filter coefficients with itself.

Consider filtering a sequence x_n to obtain a sequence y_n. Denote the operation by

$$y_n = \tilde{H}x_n$$

Suppose that I (the identity) denotes the operation

$$x_n = Ix_n$$

Then the transfer function for I is 1, where, as usual, we use $\tilde{H}(f)$.

In *Exploratory Data Analysis*, J. W. Tukey [T, Chapter 16] proposed to run the data through a filter, take the residuals

$$\epsilon_n = x_n - y_n = (I - \tilde{H})x_n$$

[where $\tilde{H}$ is the operation of applying the filter with transfer function $\tilde{H}(f)$], add these to the original signal, and then run the sum through the filter again, giving the operation

$$\tilde{H}[I + (I - \tilde{H})] = \tilde{H}(2I - \tilde{H})$$

This operator equation means that each output value of the filter $\tilde{H}$ is subtracted from $2x_n$, and the difference is then processed again by the filter $\tilde{H}$. This process Tukey calls "twicing." He also proposes further elaborations of the idea, but they will not be discussed here.

Let us examine the process more closely. For notation, we will label the original filter $\tilde{H}_{in}(f)$ and the resultant filter, after all the processing, $\tilde{H}_{out}(f)$. We now plot the *amplitude change* function $\tilde{H}_{out}(f)$ versus $\tilde{H}_{in}(f)$ (see Fig. 6.6-1). We see that when $\tilde{H}_{in}(f)$ has values near one, $\tilde{H}_{out}(f)$ has values very much closer to one; but when $\tilde{H}_{in}(f)$ is near zero, $\tilde{H}_{out}(f)$ is approximately twice as far from zero. Thus the errors in the pass band (bands) are approximately squared, whereas those in the stop band (bands) are almost doubled.

On the other hand, the opposite is true for the double use of the same filter

$$\tilde{H}_{out}(f) = \tilde{H}_{in}^2(f)$$

It is the stop part that is squared and the pass that is doubled (Fig. 6.6-1).

These two special cases immediately suggest the following approach, which will (approximately) square *small* errors in both the stop and pass bands of the filter. We want a curve of the amplitude change to be tangent

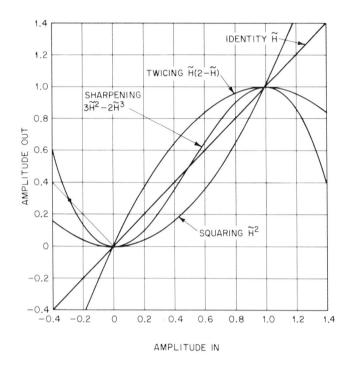

FIGURE 6.6-1   AMPLITUDE TRANSFORMATION FUNCTIONS

horizontally at both zero and one. The polynomial having this property will be a cubic of the form

$$P(X) = p_0 + p_1 X + p_2 X^2 + p_3 X^3$$

Applying the conditions on this function at both ends, we find that we have

$$P(X) = 3X^2 - 2X^3 = X^2(3 - 2X)$$

Hence we want to use

$$\tilde{H}_{\text{out}}(f) = \tilde{H}_{\text{in}}^2(f)[3I - 2\tilde{H}_{\text{in}}(f)]$$

Three passes through the same filter result in a great sharpening up of the filter. The small deviations from zero and one are both squared, and the transition band (bands) width is left the same. This situation applies to lowpass, highpass, and bandpass filters of any complexity.

So if we have a program that does some filtering job fairly well and convert it to a subroutine, then a single, short program to (a) process the signal once, (b) double this output (c) subtract each value from $3x_n$, and (d) finally pass this difference through the filter twice more will give a greatly sharpened filter of approximately three times the effective length. Of course, by suitably convolving the coefficients of the original filter, we could construct the equivalent filter and then process the signal only once. Figure 6.6-1 gives the amplitude change function. Notice that for negative values of $\tilde{H}_{in}$ and for values greater than one, the $\tilde{H}_{out}$ produces values such that $0 \leq \tilde{H}_{out} \leq 1$. Thus the method squares fairly good filters. But also notice that poor filters may be made worse, since for $\tilde{H}_{in}$ either $-0.281\ldots$ or $1.281\ldots$ gives $\tilde{H}_{out}$ the same magnitude of error, and beyond this range the values out are worse than the values in.

A study of Fig. 6.6-2 shows that a poor filter (smoothing by 3s, Section 3.2) is made worse in some places by sharpening. Figure 6.6-3 shows a slight improvement for smoothing by 5s. Moreover, Fig. 6.6-4 shows that smoothing by *both* 3s and 5s is a good filter and that "sharpening" it greatly improves the filter. The smoothing by 3s and 5s is simply the convolution of

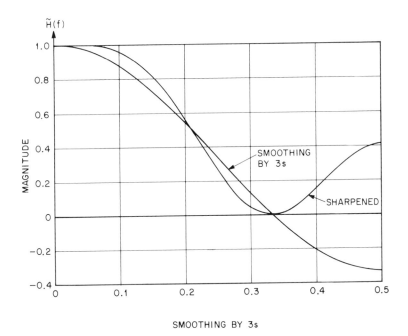

SMOOTHING BY 3s

FIGURE 6.6-2 SMOOTHING BY 3S SHARPENED

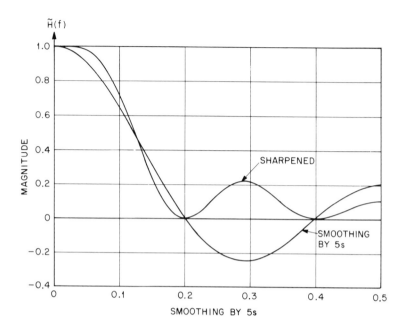

FIGURE 6.6-3 Smoothing by 5s Sharpened

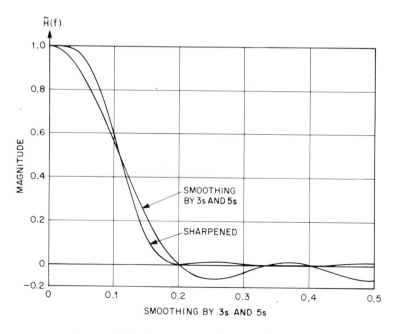

FIGURE 6.6-4 Smoothing by 3s and 5s Sharpened

three consecutive 1s with five consecutive 1s (all divided by 3 × 5), which is the filter

$$\frac{1}{15}[1, 2, 3, 3, 3, 2, 1]$$

Chapter 8 will show that if we tried to design a filter of the quality of the combination, we would find (for reasonably good filters) that we could design one of approximately double the length of the original filter. The procedure, of course, requires complete redesigning of the filter, as well as extra programming. Again, if the filter were on an integrated-circuit chip, then the proper use of three such chips would probably be cheaper than the redesign and construction of the new chip. In any case, the concept of the amplitude transformation process and its corresponding picture sheds much light on the general area of the combination of copies of the same or even different filters.

Exercises

6.6-1 Using the methods of this section, show that if we require second-order tangency at the two ends of the interval, then the corresponding amplitude change function is

$$H^3[10 - 15H + 6H^2]$$

The corresponding points of no improvement are −0.264 . . . and 1.264 . . . .

6.6-2 Show that when we require nth-order tangency at the two ends, the amplitude change function has the form

$$H^{n+1} \sum_{k=0}^{k=n} \frac{(n + k)!}{k!\, n!}(1 - H)^k$$

7

Smooth
Nonrecursive Filters

7.1 OBJECTIONS TO RIPPLES IN A
TRANSFER FUNCTION

The filters designed in the previous chapter had ripples in the transfer function, and, for some purposes, this situation is undesirable. For instance, ripples in the pass band are objectionable in cases in which the filtering is cascaded, one filter after another. If the signal passes through M identical filters, then any peak of $1 + \delta$ becomes

$$(1 + \delta)^M$$

which can cause overload (overflow in a fixed-point computer) for sufficiently large M. In sophisticated signal processing the cascading of filters is not unusual (for example, in long-distance telephone communication).

Again, if the problem is one in which the signal (information) is masked by a large amount of noise, then after filtering out the noise, any small peaks that are left in the spectrum of the signal might be from the original signal or from ripples in the transfer function used in the filtering process. Careful analysis could separate the two, but the problem can also be avoided if a class of filters that vary smoothly rather than ripple is used. By "vary smoothly" we mean that the filters are *monotone* over long intervals of the frequency band.

An example of a monotone filter occurs when we interpolate midpoints between some equally spaced data. Such problems arise in many places. For instance, in demography data is sometimes published as applying for the year end and sometimes for the midyear; in order to use data from two such different sources, it is necessary to interpolate midpoints in one of the sets of data, and one would like, typically, a monotone interpolation value filter.

Assuming the usual polynomial approximation, we get for linear interpolation for the midpoint

$$y_{n+1/2} = \frac{1}{2}[x_{n+1} + x_n]$$

If we normalize our problem and imagine that we are interpolating at the origin, then we have the two-term formula

$$y_0 = \frac{1}{2}[x_{1/2} + x_{-1/2}]$$

which we write in the symbolic form

$$\left(\frac{1}{2}\right)[1, 1]$$

Passing third-, fifth-, and seventh-degree polynomials through four, six, and eight terms, respectively, gives the following symbolic formulas:

$$\left(\frac{1}{16}\right)[-1, 9, 9, -1]$$

$$\left(\frac{1}{256}\right)[3, -25, 150, 150, -25, 3]$$

$$\left(\frac{1}{2048}\right)[-5, 49, -245, 1225, 1225, -245, 49, -5]$$

Their corresponding transfer functions are shown in Fig. 7.1-1. We see that, as expected, the higher-order polynomial has the higher-order tangency at the origin and comes close to being an all-pass filter.

If we design a filter that removes more of the high frequency, we will find that making it tangent at the folding frequency $f = \frac{1}{2}$ as well as at the origin will, for low-order filters, give all positive coefficients. This situation reminds us that

THEOREM.   *Having all the coefficients positive is the necessary and sufficient condition for a smoothing filter also to have the property that if one function is uniformly larger than a second, then its output will be larger as well.*

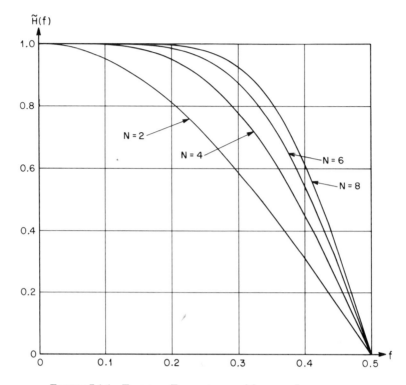

FIGURE 7.1-1 TRANSFER FUNCTIONS FOR MIDPOINT INTERPOLATION

But for such a filter this statement means that if there is a local peak, the interpolated value must be less than the maximum of the values (it is a weighted average with positive weights); thus peaks are cut off (and valleys are filled in). The peaks and valleys can be reasonably accurately portrayed by the filter only if we allow high frequencies to pass and thereby allow positive and negative coefficients in the filter.

Here is one of the many difficulties of filtering: we simply cannot have everything that we want. By removing the noise, we tend to force the response to changes in the signal to be less—we tend to flatten out peaks and valleys and other sudden changes. If we insist on keeping the small details of the data, then we must also keep the high-frequency noise.

Exercises

7.1-1   Prove the theorem in this section.

7.1-2   Derive the formulas for polynomial midpoint interpolation.

7.2   SMOOTH FILTERS

Let us once again review where we are in the topic of filters. Given a symmetric ($c_k = c_{-k}$), nonrecursive digital filter operating on equally spaced data, x_n, from some source, we compute a y_n by the formula

$$y_n = \sum_{k=-N}^{k=N} c_k x_{n-k}$$

The eigenfunctions of linear problems are the complex exponentials $e^{2\pi i f n}$. The use of the eigenfunctions leads to the transfer function (eigenvalues)

$$\tilde{H}(f) = c_0 + 2 \sum_{k=1}^{k=N} c_k \cos 2\pi k f$$

We now show the well-known fact that

$$\cos k\theta = \text{polynomial in } \cos \theta \text{ of degree } k$$

To prove it, we begin with the simple observation that

$$e^{in\theta} = [e^{i\theta}]^n$$

Next, write this equation in the complex form

$$\cos n\theta + i \sin n\theta = [\cos \theta + i \sin \theta]^n$$

expand the binomial, and take the real part

$$\cos n\theta = \sum_{k=0}^{n} C(n, 2k) \cos^{n-2k} \theta (i \sin \theta)^{2k}$$

where, of course, the summation is cut off when $2k$ exceeds n, because then the binomial coefficients are zero. Since

$$\sin^{2k} \theta = [\sin^2 \theta]^k = [1 - \cos^2 \theta]^k$$

we have the desired polynomial in powers of $\cos \theta$.

Returning to our transfer function

$$\tilde{H}(f) = c_0 + 2 \sum_{k=1}^{N} c_k \cos 2\pi k f$$

we can use the preceding result to get, for suitable b_k's,

$$\tilde{H}(f) = \sum_{k=0}^{N} b_k [\cos 2\pi f]^k$$

We now make the transformation of the independent variable

$$\cos 2\pi f = t$$

As f goes from 0 to $\frac{1}{2}$, t goes from 1 to -1, and we will have the polynomial in t

$$\tilde{H}(f) = \sum_{k=0}^{N} b_k t^k$$

as an equivalent transfer function. However, note that the transformation is a nonlinear stretching of the frequency axis. By working in the variable t, which is $\cos 2\pi f$, we will be representing the transfer function in terms of powers of $\cos 2\pi f$ rather than in cosines of multiple angles as we have been doing up to now. The original lowpass filter, because of the reversal of the axis in the transformation, now "looks like" a highpass filter in the t variable.

To get started on the design, we choose the function

$$g(t) = [1 + t]^p [1 - t]^q$$

with p and q as parameters (Fig. 7.2-1). Clearly, this function has a zero of order p at $t = -1$ and a zero of order q at $t = 1$. Integrating this function of t gives an arbitrary constant of integration, which we will fix so that at

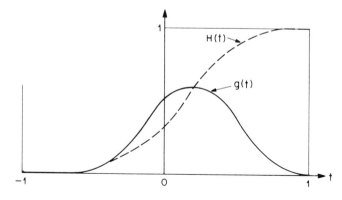

FIGURE 7.2-1

$t = -1$ the integrated function is zero. Next, we will calculate the value of the function at $t = 1$ and divide by this number (normalize) so that the final function has the value 1 at $t = 1$ (Fig. 7.2-1).

$$H(t) = \frac{\int_{-1}^{t} (1 + t)^p (1 - t)^q \, dt}{\int_{-1}^{1} (1 + t)^p (1 - t)^q \, dt}$$

Those familiar with the function may recognize it as the incomplete Beta function in disguise; the normalizing factor is the corresponding complete Beta function. After integration we have a polynomial in t (our transfer function) that has a $(p + 1)$th-order zero at $t = -1$ and the value 1 at $t = 1$, along with q derivatives that are zero at $t = 1$. The reason that the order of the highest vanishing derivative rises by 1 is that the process of integration increases the tangency at the origin and at the terminal value.

This function in t has the coefficients b_k in the transfer function (no relation to the Fourier coefficients b_k), and the simple substitution of $t = \cos 2\pi f$ will bring us back to the frequency variable. To follow this transformation, we have

$$\tilde{H}(f) = H(\cos 2\pi f) = \frac{\int_{-1}^{\cos 2\pi f} (1 + t)^p (1 - t)^q \, dt}{\int_{-1}^{1} (1 + t)^p (1 - t)^q \, dt}$$

The denominator is a constant; thus we need only examine the behavior of the numerator. We make the transformation in the integral

$$t = \cos 2\pi f$$

Therefore, using the half-angle formulas,

$$\int_{-1}^{\cos 2\pi f} = \int_{1/2}^{f} (1 + \cos 2\pi f)^p (1 - \cos 2\pi f)^q (-2\pi \sin 2\pi f) \, df$$

$$= 2^{p+q+2} \pi \int_{f}^{1/2} (\cos \pi f)^{2p+1} (\sin \pi f)^{2q+1} \, df$$

$$= 2^{p+q+2} \pi \int_{f}^{1/2} \left[\sin \pi \left(\frac{1}{2} - f \right) \right]^{2p+1} \left[\sin \pi f \right]^{2q+1} df$$

Since $\sin \pi f$ is approximately πf, near a zero of a sine function we see the doubling in the order of tangency at both ends of the interval due to the "stretching out" caused by the transformation.

A simple method, which will be developed in the next section, enables us to make the transformation back to the Fourier series representation of the transfer function in order to get the c_k coefficients of the digital filter that we started out to find.

Returning to the design problem, we equate $g'(t)$ to zero and find that the inflection point of the function $H(t)$ in the t domain occurs at

$$\frac{p - q}{p + q}$$

Increasing both parameters p and q proportionately gives a narrower transition zone in $H(t)$.

The original curve $g(t)$ is a polynomial in t with all its features assigned at the ends of the interval; thus there are no "wiggles" in both the $g(t)$ and $H(t)$ curves between the endpoints. The transformation back to the f variable is monotone and only stretches the independent variable axis, which means that it cannot put in maxima and minima. The flatness of the cosine curve of the transformation doubles the tangencies at the ends.

We have yet to do the transformation back to the Fourier series notation, but this step is only a change of notation and does not affect the shape of the curve. Thus we obtain a smooth transfer function. The direct design method using the Fourier series terms of the form $\cos 2\pi kf$ tends to produce wiggles in the transfer function. It is to escape these wiggles that we have made the fit very good at the ends of the interval and let the rest fall where it may, using only the degrees of the tangencies at the ends to control where the filter cutoff occurs.

Exercises

7.2-1   Derive the formula for the location of the inflection point.

7.2-2   Convert $1 + t + t^2 + t^3$ to a Fourier series, where $t = \cos 2\pi f$.

7.3   TRANSFORMING TO THE FOURIER SERIES

The problem discussed in this section is how to get from a power series in $\cos \theta$ to a Fourier series in $\cos k\theta$. Furthermore, we want to be able to do it easily on a computing machine. Thus we look for a recursive way of doing the conversion that will make the programming easy.

Let the power series in $\cos \theta$

$$\sum_{k=0}^{N} b_k [\cos \theta]^k$$

be written in the usual "chain" form.

$$b_0 + \{ \cdots + \cos \theta \, [b_{N-2} + \cos \theta \, (b_{N-1} + b_N \cos \theta)] \cdots \}$$

Starting at the inner parenthesis with the last two terms

$$b_{N-1} + b_N \cos \theta$$

we multiply by $\cos \theta$ and add the next lower coefficient. Then we again multiply by $\cos \theta$ and add the next lower coefficient and so on.

The first two terms are in the form of a Fourier series and thereby form the basis for an induction (a recursive process). Consequently, we assume that at each stage we have a Fourier series with the coefficients given and show that the next stage is also a Fourier series. To show this we multiply the current Fourier series in the induction process by $\cos \theta$ and use the formula for the product of two cosines

$$\cos \theta \cos n\theta = \frac{1}{2} [\cos (n + 1)\theta + \cos (n - 1)\theta]$$

From any one coefficient with index greater than zero, we get two terms—the first is one unit in frequency higher and the second is one frequency lower, both with the coefficient $\frac{1}{2}$. The constant term is, of course, now a cosine term with the full coefficient. Finally, the next lower coefficient b_k is added to the constant term that came from the earlier calculations of the term with index equal to one. See Fig. 7.3-1, where the dots mean the corresponding coefficients of the Fourier terms of increasing frequency toward the right and the small figures alongside the arrows indicate the multipliers. If this simple process is repeated often enough, we end up with the Fourier series that is equivalent to the original power series in $\cos \theta$. The coefficients in the corresponding complex Fourier series are the coefficients c_k of the filter.

Notice that the process has a number of nice properties. First, it is robust; the division by 2 at the various stages produces no roundoff in floating-point binary arithmetic and tends to decrease errors. Second, it is simple to program, and involves no tables and little arithmetic beyond additions. Finally, this same process will be used in another situation; thus it should be under-

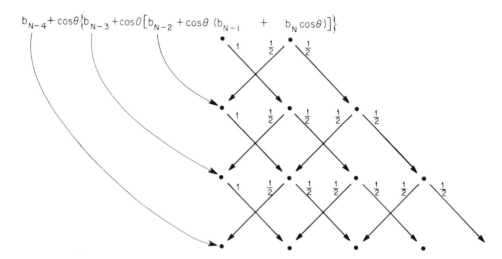

$$b_{N-4} + \cos\theta\{b_{N-3} + \cos\theta[b_{N-2} + \cos\theta\,(b_{N-1} \quad + \quad b_N\cos\theta)]\}$$

FIGURE 7.3-1 ONE STAGE IN TRANSFORMING TO THE FOURIER SERIES

stood. Evidently it depends on the simple trigonometric formula for the product of two cosines and nothing more beyond the organization of the computation into a regular form.

Exercises

7.3-1   Convert $\cos^5\theta$ to an equivalent Fourier series.

7.3-2   Convert $\cos^4\theta + \cos^2\theta + \cos\theta + 1$ to an equivalent Fourier series.

7.4   POLYNOMIAL PROCESSING IN GENERAL

While on the general topic, which might be called "the engineering of algebra," we will pause to organize the handling of polynomials that our design process requires. Again, we will see that recursive methods are preferable to the ones that might come to mind as a result of taking the usual mathematics courses. Generally such courses concentrate on ideas and minimize the operations of carrying them out. It is clearly necessary to understand what we are doing before mastering the way to do it; so we are not faulting mathematics courses by saying that they neglect the engineering of algebra. Yet in order to do extensive symbol manipulation, we cannot overlook the situation. We shall organize matters both for the computer and for small designs done by hand.

How, for example, do we do some of the algebra required in the preceding design? We began with the expression

$$[1 + t]^p[1 - t]^q$$

and obtained a polynomial in t. Let us suppose that $p \leq q$. It is relatively easy to handle the opposite case. We write

$$q = p + k \qquad (k \geq 0)$$

Therefore we have

$$[1 - t^2]^p[1 - t]^k$$

The first square bracket can be expanded by the standard binomial process to get the series in powers of t^2. Notice that (a) viewed as a polynomial in t, every other term is zero; (b) by the usual process, the binomial coefficients are recursively calculated; and (c) a zero indicates the end of the recursion. The maximum binomial coefficient is the middle one, but it refers to index p, whereas we have an expansion in t up to powers $2p$. To multiply this expansion by $(1 - t)$, we need only copy the set of coefficients shifted one place to the right and then subtract from the original coefficients. The first step merely combines nonzero numbers with the zeros and actually does no arithmetic but fits into the recursion. This process of shifting the set of coefficients one to the right and then subtracting should be done exactly k times in order to allow for the factor $(1 - t)^k$. It can be done "in place" in a computer.

To integrate the resulting polynomial, we merely divide the kth coefficient by $k + 1$ (and, of course, increase in our minds the power of t that it represents). Evaluating this polynomial at $t = -1$ consists of simply adding the coefficients with alternating sign and determining the constant of integration term C that is needed to make the integral zero at $t = -1$.

Up to now we can use any (nonzero) multiple of the polynomial that we please, since the final normalization to make the polynomial have the value 1 at $t = 1$ will remove this multiplying factor.

Now that we have the constant term we can evaluate the polynomial at $t = 1$. To do so, we merely sum all the coefficients. Finally, dividing all the coefficients by this number normalizes the integral so that it has the value 1 at $t = 1$. Thus we have our transfer function in polynomial form.

Again, we see that simple organization based on recursive methods makes larger masses of algebra practical, both on a machine and by hand.

7.5 THE DESIGN OF A SMOOTH FILTER

Suppose that we want to design a smooth filter that passes the lower one-third of the Nyquist interval—that is, up to $\pi/3$—and stops frequencies in the upper third. Going to the t coordinate, the edge at the filter occurs at

$$\cos\left(\frac{\pi}{3}\right) = t = \frac{1}{2}$$

So we need to select p and q such that (at least approximately)

$$\frac{p - q}{p + q} = \frac{1}{2}$$

The two choices $p = 3$, $q = 1$ and $p = 6$, $q = 2$ differ in the sharpness of the cutoff.

Case 1. $p = 3$, $q = 1$ (almost trivial)

The starting polynomial is

$$(1 + t)^3(1 - t) = (1 - t^2)(1 + t)^2$$

Powers of $t \longrightarrow$		1	t	t^2	t^3	t^4	t^5
$1 - t^2$	$=$	1	0	-1			
			1	0	-1		
$(1 - t^2)(1 + t)$	$=$	1	1	-1	-1		
			1	1	-1	-1	
$(1 - t^2)(1 + t)^2$	$=$	1	2	0	-2	-1	
Integrate $\longrightarrow$		C	1	1	0	$-\frac{1}{2}$	$-\frac{1}{5}$
Fix constant $\longrightarrow$		$\frac{1}{10}[3$	10	10	0	-5	$-2]$
Scale		$\frac{1}{16}[3$	10	10	0	-5	$-2]$

These are the b_k's in the $\cos(2\pi f)$ expansion. To avoid fractions in the conversion to the c's of the Fourier expansion, we write the polynomial in the form (multiply and divide by $2^4 = 16$)

$$\left(\frac{1}{16}\right)^2 [48 \quad 160 \quad 160 \quad 0 \quad -80 \quad -32]$$

The conversion process from the b_k's to the Fourier coefficients a_k's is

$$
\begin{array}{ccccccc}
-80 & -32 \\
\hline
-16 & -80 & -16 \\
0 \\
\hline
-16 & -80 & -16 \\
-40 & -8 & -40 & -8 \\
160 & -16 \\
\hline
120 & -24 & -40 & -8 \\
160 & -20 & -4 & -20 & -4 \\
-12 & 120 & -12 \\
\hline
148 & 100 & -16 & -20 & -4 \\
48 & 148 & -10 & -2 & -10 & -2 \\
50 & -8 & 50 & -8 \\
\hline
98 & 140 & 40 & -10 & -10 & -2 \\
\hline
\end{array}
$$

Thus the filter is

$$\left(\frac{1}{16}\right)^2 [-1 \quad -5 \quad -5 \quad 20 \quad 70 \quad 98 \quad 70 \quad 20 \quad -5 \quad -5 \quad -1]$$

As a result, we have the monotone filter shown in Fig. 7.5-1.

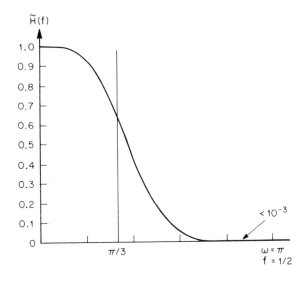

FIGURE 7.5-1 LOWPASS MONOTONE FILTER

7.6 SMOOTH PASS BAND FILTERS

To extend these ideas to a band pass filter, we need only start with a different function. Again we want zeros at both ends, but this time we also want an odd-order zero where the middle of the pass band occurs. Thus we start with a polynomial of the form [Fig. 7.6-1(a)]

$$(t + 1)^p(t - t_0)^{2q+1}(t - 1)^r$$

where we have not as yet picked the value t_0. Choice of the exponents is clear once we realize that the p and r tend to control where the pass band will be and that the $2q + 1$ controls the flatness of the pass band.

Next, we expand this polynomial first by using the binomial expansion on the middle term to get a polynomial in t_0. Clearly, a computer is needed to do all the arithmetic! Each coefficient is a polynomial in t; and using the same methods illustrated in the previous section, we arrange each coefficient of t_0 as a polynomial in t.

Integration of this polynomial from -1 to t will show that the integral will usually not be zero at $t = 1$ for our initial choice of t_0 [Fig. 7.6-1(b)].

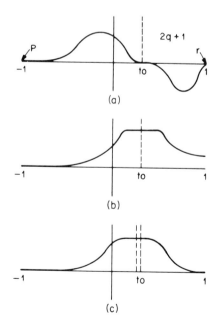

FIGURE 7.6-1

Each of the coefficients of the powers of t_0 can be evaluated at $t = 1$, and we have a polynomial in t_0 with numerical coefficients. The bisection method (or Newton's method) can be used to find the *simple* zero at t_0 such that the integral will be zero at $t = 1$. After finding this t_0, we have our integrated polynomial except that we must still normalize it so that at the value t_0 it has the value 1 [Fig. 7.6-1(c)].

Once we have the t_0 and the normalization, then the filter process design is much as before. Conversion to the Fourier series form and then to the ultimate digital filter consists of the same processes used earlier. It is clear that these designs are best done on a computer and that the actual programming is recursive and short. The actual machine time used is not high either.

8

The Fourier Integral and the Sampling Theorem

8.1 INTRODUCTION

The Fourier series is useful for handling both periodic functions and functions with a finite range of independent variables (since this finite range may be extended to the whole line by defining a periodic extension of the function). However, periodic functions are comparatively rare in practice; and in order to consider many of the applications of digital filter, a more general class of functions is necessary. This situation, in turn, requires the development of a more general mathematical tool to handle the nonperiodic functions—namely, the Fourier integral. A simple example of a nonperiodic function whose components are periodic is

$$y(t) = \cos t + \cos \sqrt{2}\, t$$

Since 1 and the $\sqrt{2}$ are not commensurate, this function cannot repeat itself exactly no matter how far we go in the variable t.

The formal definition of the Fourier integral representation for $g(t)$ is

$$g(t) = \int_{-\infty}^{\infty} G(f)e^{2\pi i f t}\, df$$

Clearly, in the Fourier integral, there are a noncountable number of frequencies, one corresponding to each real number f.

In practice, we will usually deal only with frequencies in a band, for, as we saw in Chapter 2, the process of sampling at a sequence of equally spaced points produces an effect that can be viewed as aliasing each frequency into a frequency lying in the Nyquist interval. Such functions are called *band limited* because, for practical purposes, all their frequencies are within a band such that at least two samples in the highest frequency present is the rule. Physical examples of continuous functions that are almost band limited are common; for instance, the typical hi-fi system has a cutoff at the low end of a few ten to possibly a hundred hertz, and at the upper end the cutoff frequency may be as high as 20,000 Hz.

One purpose of this chapter is to form a mathematical bridge between the original continuous function that we customarily think about and the sampled function that must be dealt with when using a digital filter. We also need to know how to go from the sampled function that is band limited back to the original continuous function—how to interpolate the function between the given equally spaced samples. For these reasons, we must look at the famous *sampling theorem*.

Finally, we examine the effect of taking a finite piece of a continuous function from what is potentially an infinitely long function and various ways of using windows to reduce this effect.

8.2 SUMMARY OF RESULTS

This section summarizes the rest of the chapter and does not try to prove the results rigorously, since rigorous proofs would draw us deep into the details of mathematics and obscure what is happening.

The first thing to show is that for any reasonable function $g(t)$ there is a representation in the form of a Fourier integral

$$g(t) = \int_{-\infty}^{\infty} G(f)e^{2\pi i f t}\, df$$

where we also have

$$G(f) = \int_{-\infty}^{\infty} g(t)e^{-2\pi i f t}\, dt$$

These formulas are sometimes called the *Fourier inversion formulas*. In analogy with the Fourier series, from which we shall derive the Fourier integral, we have the *density function* $G(f)$ corresponding loosely to the c_k of the complex form of the Fourier series (Section 5.4); the density $G(f)$ times the interval Δf gives the amount in measureable units. The function $G(f)$ is called *the transform* of $g(t)$. We shall adopt the customary usage of lowercase letter in the time domain and the corresponding capital letter in the frequency domain. As can be seen, the only difference (in the f notation but *not* in the ω notation) in the two equations is that one has a $-i$ in place of i.

Aliasing plays a fundamental role in sampling. The earlier discussion of aliasing with respect to Fourier series (Section 2.2) was in no way dependent on the discrete frequencies that occurred, and everything goes through exactly the same for the continuous range of frequencies used in the Fourier integral. If the frequencies are limited to a symmetric band about the origin

$$-F \leq f \leq F$$

of width $2F$ and the sampling is done at an interval of Δt between the samples, then we have the important relation

$$2F\,\Delta t < 1$$

as a necessary condition to avoid aliasing (the inequality rather than the equality is necessary to avoid the end frequencies that are theoretically troublesome but that are easy to cope with in practice). The maximum sampling interval $1/(2F)$ is related to the folding or Nyquist frequency. The common way of expressing this relationship is to say, as before, that we must have at least two samples in the highest frequency present in order to avoid aliasing.

Exercises

8.1-1   Write the Fourier integral in the angular frequency notation. How should the coefficient 2π be handled?

8.1-2   If there are a hundred samples per second, what is the folding frequency?

8.3   THE SAMPLING THEOREM

One question that naturally occurs is: Is it possible to reconstruct a band-limited function from its samples, provided that we observe the sampling restriction? The fundamental result that it *is* possible is called *the sampling theorem*, which is so important that two different, partially rigorous proofs will be given in the hope that doing so will make the theorem understandable.

As background, let us look at the corresponding result in polynomial interpolation, since interpolation is what we are doing in the sampling theorem. From the infinite number of discrete samples, we try to reconstruct the value of the band-limited function at any given point of the function. In the polynomial case, we use the Lagrange interpolation formula. The standard notation is

$$\pi(t) = (t - t_1)(t - t_2) \cdots (t - t_n) \qquad \pi_k(t) = \frac{\pi(t)}{t - t_k}$$

where $\pi_k(t)$ is the product of all the differences *except* the kth one. Now consider the expression

$$\frac{\pi_k(t)}{\pi_k(t_k)}$$

This expression is zero at all the sample points t_j *except* the kth one, at which it is exactly 1. Once we understand this property, it is easy to see that the function

$$g(t) = \sum_{k=1}^{n} g(k) \frac{\pi_k(t)}{\pi_k(t_k)}$$

has the value $g(k)$ at the kth point and is therefore the interpolating polynomial of degree $n - 1$ that passes through the given values $g(k)$.

If the frequency approach is used instead of the polynomial approach, we are led to consider the corresponding function (for unit spacing)

$$\frac{\sin \pi(k - t)}{\pi(k - t)}$$

which for $t = k$ takes on the value one and for all other integers has the value zero. Thus the formal expression

$$g(t) = \sum_{k=-\infty}^{\infty} g(k) \frac{\sin \pi(k - t)}{\pi(k - t)}$$

clearly passes through the sample values $g(k)$. We say "formal" because we have no indication at present that it will converge except at the sample points.

The original function from which we obtained the samples $g(k)$ was assumed to be band limited; is the above function $g(t)$ also band limited? It is easy to show that it is by direct integration. Taking unit spacing for our samples, the particular band-limited function that is one in the band $-\frac{1}{2}$ to $\frac{1}{2}$ and zero outside leads to

$$\int_{-1/2}^{1/2} e^{2\pi i f t}\, df = \frac{e^{\pi i t} - e^{-\pi i t}}{2i\pi t} = \frac{\sin \pi t}{\pi t}$$

Therefore the function on the right-hand side of this equation is band limited. Shifting the independent variable t a fixed amount k does not change the frequencies in the function, and we conclude (from the band-limited property of the individual terms of the formal sum) that the sum itself is band limited. This is our first, rather formal proof of the sampling theorem; from the samples $g(k)$ of a band-limited function we can reconstruct the original function. Some attention to the uniqueness of the function is, of course, necessary.

In analogy with the power spectrum in the Fourier series, we have the quantity

$$|G(f)|^2 = G(f)\bar{G}(f)$$

as the power spectrum in the theory of Fourier integrals. Often the word "power" is dropped, and the quantity is merely referred to as the *spectrum*.

In order to obtain a feeling for what the Fourier integral is, we make the loose analogy of $g(t)$ as a light beam. The transform, like a glass prism, breaks up the function into its component frequencies f, each of intensity $G(f)$. In optics, the various frequencies are called colors: we get via the Fourier integral the color spectrum of the incoming signal. If the input were a single frequency, then we get a *spectral line*. In practice, of course, an absolutely pure spectral line does not occur, but some are so close to a single frequency that the difference is unimportant.

Unfortunately, if the function $g(t)$ does not approach zero as $|t|$ becomes infinite, then the transform $G(f)$ does not exist in a mathematical sense. This situation is a nuisance in the common case of

$$g(t) = \cos \omega_0 t$$

especially for $\omega_0 = 0$.

The difficulty is, however, a clear warning that the model being used is unrealistic in the following sense. We can replace $g(t)$ by

$$g_1(t) = \begin{cases} \cos \omega_0 t, & |t| < T \\ 0, & |t| > T \end{cases}$$

for some suitably large T and then carry out the desired work (involving the extra details due to the ends). If the discontinuities at the ends are annoying, we can put $g_1(t)$ through the Lanczos convolving window and obtain a new $g_2(t)$ that has continuity and hence a more rapidly converging Fourier series. Indeed, any degree of smoothness at the ends can be used at the cost of more details in processing the function. We can then carry out the process for an arbitrary T. If, on letting $T \longrightarrow \infty$, we obtain peculiar results, then consider the following observation. These finite functions have finite energy. The fact that their energy becomes infinite as T becomes infinite shows the unrealistic nature of the model that assumes that they describe reality. A pure sinusoid lasting infinitely long is unrealistic, and some suitably truncated function should be practical in almost all situations.

The rest of the chapter simply describes some of the details of the preceding outline, and, as noted, we do not intend to be too rigorous. We will suppose that the functions used to represent the subject being studied are sufficiently "well behaved" to meet any special restraints on the class of functions for which the results apply and will exclude the "pathological cases." For those whose mathematical experience with complex numbers is limited, it is suggested that the subject be reviewed before continuing; otherwise the discussion will be obscured by the lack of elementary manipulative skills.

8.4 THE FOURIER INTEGRAL

We now derive the Fourier integral from the Fourier series. Given the Fourier series of a function that is periodic in the interval $-N < t \leq N$ (Section 5.4), we can eliminate the coefficient c_k in the complex form and obtain

$$g(t) = \sum_{k=-\infty}^{\infty} \left[\int_{-N}^{N} g(t') e^{-(\pi i/N)kt'} \, dt' \right] e^{(\pi i/N)kt} \frac{1}{2N}$$

In order to approach the approximation of nonperiodic functions, we let the interval get longer and longer—that is, we let $N \longrightarrow \infty$. In the limit the function is no longer periodic, since the interval of periodicity is the whole axis.

To see what happens as $N \longrightarrow \infty$, we set

$$\frac{1}{2N} = \Delta f$$

and hence

$$\frac{k}{2N} = k\,\Delta f = f_k$$

Our equation then becomes ($k = 2Nf_k$)

$$g(t) = \sum_{k=-\infty}^{\infty} \left[\int_{-N}^{N} g(t')e^{-2\pi i f_k t'}\,dt' \right] e^{2\pi i f_k t}\,\Delta f_k$$

It is easy to see that as N grows larger and larger, the successive f_k in the summation get closer and closer to each other; the exponentials are being pushed closer together, and the sum is approaching an integral. It is reasonable to suppose that in the limit the summation becomes an integral, *provided* that the function $g(t)$ is reasonably well behaved. In the limit the equation becomes

$$g(t) = \int_{-\infty}^{\infty} \left[\int_{-\infty}^{\infty} g(t')e^{-2\pi i f t'}\,dt' \right] e^{2\pi i f t}\,df$$

To get the Fourier integral in the usual form, we set

$$G(f) = \int_{-\infty}^{\infty} g(t')e^{-2\pi i f t'}\,dt'$$

and we have

$$g(t) = \int_{-\infty}^{\infty} G(f)e^{2\pi i f t}\,dt$$

The use of the rotational frequency f rather than the angular frequency ω produces a nice symmetric representation without any extra numerical coefficients. The function $G(f)$ is said to be *the Fourier transform* of the function $g(t)$. The two functions have almost exactly the same relationship to each other; the exception is that in the exponent one integral has i and the other $-i$. Both functions contain the same information in the sense that each can be found from the other; but they exhibit the information in significantly different forms. It is the power of these alternate forms that makes the Fourier transform so useful in understanding what is occurring in many situations.

8.5   SOME TRANSFORM PAIRS

The importance of the relationship between the two functions $g(t)$ and $G(f)$ suggests that a table transforms would be useful. Such tables are widely available. For our immediate use, we will develop only a few such relationships that we will need, and leave the more extensive development of the Fourier integral to books that specialize in the subject.

Our first example of a Fourier integral transform is the band-limited function with unit area [Fig. 8.5-1(a)]

$$G(f) = \begin{cases} \dfrac{1}{2f_c}, & \text{for } |f| < f_c \\ 0, & \text{for } |f| > f_c \end{cases}$$

We now set up the integral for $g(t)$ and, as in Section 8.3, obtain

$$g(t) = \int_{-\infty}^{\infty} G(f)e^{2\pi i f t}\, df = \frac{1}{2f_c}\int_{-f_c}^{f_c} e^{2\pi i f t}\, dt$$

$$= \frac{1}{2f_c}\frac{e^{2\pi i f t}}{2\pi i t}\bigg|_{-f_c}^{f_c} = \frac{e^{2\pi i f_c t} - e^{-2\pi i f_c t}}{2i}\frac{1}{2f_c\pi t}$$

$$= \frac{\sin 2\pi f_c t}{2\pi f_c t}$$

This function is very common in the theory, and an idea of its behavior around the origin is given by its power series expansion

$$\frac{\sin 2\pi f_c t}{2\pi f_c t} = 1 - \frac{(2\pi f_c t)^2}{3!} + \frac{(2\pi f_c t)^4}{5!} - \cdots$$

Figure 8.5-1(b) gives the graph of the function. The main lobe extends from $-1/(2f_c)$ to $1/(2f_c)$; beyond that the function oscillates at the constant rate, while the amplitude gradually dies out like $1/t$. The larger the f_c, the narrower the central peak and the oscillations.

Our second example of a Fourier transform is actually a relationship between Fourier transforms. Suppose that we already know a pair of Fourier transforms $g_1(t)$ and $G_1(f)$. Thus we assume that we know

$$g_1(t) = \int_{-\infty}^{\infty} G_1(f)e^{2\pi i f t}\, df$$

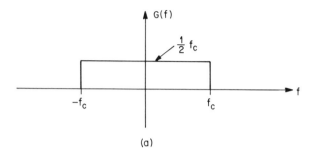

(a)

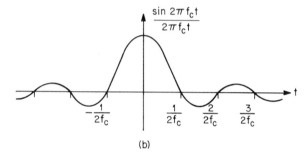

(b)

FIGURE 8.5-1

We now ask: What function $g_2(t)$ corresponds to $G_1(f)e^{2\pi i f y}$? From the definition, we have

$$g_2(t) = \int_{-\infty}^{\infty} [G_1(f)e^{2\pi i f y}]e^{2\pi i f t}\, df$$

$$= \int_{-\infty}^{\infty} G_2(f)e^{2\pi i f(y+t)}\, df$$

Clearly, $y + t$ plays the role of t in the original function, so that

$$g_2(t) = g_1(y + t)$$

The effect of the exponential multiplier is to shift the argument of the transformed function. This result is often called the *shifting theorem*.

Exercises

8.5-1   If $g(t) = (1/\sqrt{2\pi})e^{-at^2}$, find its Fourier transform.

8.5-2   Find the Fourier transform of $e^{-a|t|}$.

8.6 BAND-LIMITED FUNCTIONS AND THE SAMPLING THEOREM

Because of the importance of the sampling theorem in computing, we present a second, slightly more rigorous proof in this section. We are still avoiding pathological functions and excessive rigor.

The central idea of the sampling theorem is that a band-limited function $g(t)$ extending from $t = -\infty$ to $t = \infty$ is sampled at equally spaced points with a spacing such that at least two samples per cycle occur in the highest frequency present. As noted, the restriction to band-limited functions corresponds to a natural physical limitation in many situations. In any case, it is forced on us if we sample a function, since the very act of sampling produces the aliasing of the higher frequencies into the frequencies within the Nyquist band. However, a word of caution: the mathematical model being used asserts that if a signal is band limited, then it cannot be time limited; and, conversely, if it is time limited, then it cannot be band limited. In practice, no signal lasts forever, and so in the mathematical model it cannot be band limited. Evidently the mathematical model should not be taken too literally when applying results to the real world; it is a useful, not a sacrosanct model.

To derive the sampling theorem, suppose that we are given the band-limited function $G(f)$, which is zero for $|f| > f_c$. The first step in the derivation is to replace this function with a periodic function (in order to use our theory of Fourier series) that agrees with $G(f)$ in the band; thus we define the function $G_1(f)$, which coincides with $G(f)$ inside the interval $|f| < f_c$ and is periodic outside, as shown in Fig. 8.6-1.

Using the Fourier integral, we have

$$G(f) = \int_{-\infty}^{\infty} g(t)e^{-2\pi i f t}\, dt$$

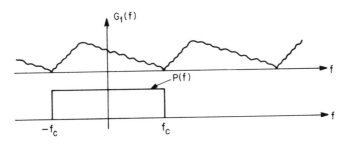

FIGURE 8.6-1

Since $G(f)$ is band limited, and in the band we have $G(f) \equiv G_1(f)$,

$$g(t) = \int_{-\infty}^{\infty} G(f)e^{2\pi i f t}\, dt = \int_{-f_c}^{f_c} G(f)e^{2\pi i f t}\, dt = \int_{-f_c}^{f_c} G_1(f)e^{2\pi i f t}\, df$$

For $G_1(f)$ we have, since we made it periodic, the Fourier series expansion

$$G_1(f) = \sum_{k=-\infty}^{\infty} c_k e^{(\pi i/f_c)kf}$$

where

$$c_k = \frac{1}{2f_c}\int_{-f_c}^{f_c} G_1(f)e^{-(\pi i/f_c)kf}\, df$$

But from the equation for $g(t)$ this integral is the same as

$$c_k = \frac{1}{2f_c}g\left(\frac{-k}{2f_c}\right)$$

In order to "cut out" the original function $G(f)$ from the new function $G_1(f)$, we multiply $G_1(f)$ by the rectangular function $2f_c P(f)$, where (see Fig. 8.5-1 again)

$$P(f) = \begin{cases} \dfrac{1}{2f_c}, & \text{for } |f| < f_c \\[2mm] 0, & \text{for } |f| > f_c \end{cases}$$

We have already found that the transform of $P(f)$ is the band-limited function

$$p(t) = \frac{\sin 2\pi f_c t}{2\pi f_c t}$$

Thus

$$G(f) = G_1(f)P(f)2f_c$$

For $G_1(f)$ and the c_k, we substitute the results just found, and we have

$$G(f) = \sum_{k=-\infty}^{\infty} g\left(\frac{-k}{2f_c}\right)P(f)e^{(\pi i/f_c)kf}$$

The next step is to transform the results back to the time domain via the Fourier transform and apply the shifting theorem

$$g(t) = \int_{-\infty}^{\infty} G(f)e^{2\pi i f t}\, df$$

$$= \sum_{k=-\infty}^{\infty} g\left(\frac{-k}{2f_c}\right) \int_{-\infty}^{\infty} P(f)e^{\pi i k/f_c}e^{2\pi i f t}\, df$$

$$= \sum_{k=-\infty}^{\infty} g\left(\frac{-k}{2f_c}\right) \frac{\sin 2\pi f_c(t + k/2f_c)}{2\pi f_c(t + k/2f_c)}$$

Replacing k by $-k$ as the dummy summation index, we have the sampling theorem

$$g(t) = \sum_{k=-\infty}^{\infty} g\left(\frac{k}{2f_c}\right) \frac{\sin \pi(2f_c t - k)}{\pi(2f_c t - k)}$$

At the folding frequency itself we cannot reconstruct the function from the samples because, for unit spacing, the function $g(t) = \sin \pi t$ is identically zero at all the sample points and the sampling theorem would therefore give the identically zero function. Thus in the "splitting hairs" mathematical approach, we must have *more* than two samples per cycle for the highest frequency present.

8.7   THE CONVOLUTION THEOREM

For our purposes, perhaps the most important theorem in the theory of Fourier integrals (after the inversion formula) is the *convolution theorem,* which corresponds to the earlier convolution theorems for Fourier series (Section 5.5). Suppose that we have two functions, $g_1(t)$ and $g_2(t)$. The convolution $h(t)$ of $g_1(t)$ with $g_2(t)$ is defined as

$$h(t) = \int_{-\infty}^{\infty} g_1(s)g_2(t - s)\, ds$$

(Notice how it differs from the earlier definition in Section 5.5, where we assumed periodicity of the functions.) The variable of integration is s; and regarding t as fixed with respect to the integration, we make the change of variable of integration

$$t - s = s'$$

The convolution becomes

$$h(t) = \int_{-\infty}^{\infty} g_1(t - s')g_2(s')\, ds'$$

So (as before) the convolution of $g_1(t)$ with $g_2(t)$ is the same as the convolution of $g_2(t)$ with $g_1(t)$.

It is natural to ask: What is the Fourier transform of the convolution? By definition, the Fourier transform of $h(t)$ is

$$H(f) = \int_{-\infty}^{\infty} h(t)e^{-2\pi ift}\, dt$$

Substituting for $h(t)$, we get

$$H(f) = \int_{-\infty}^{\infty} g_1(s)e^{-2\pi ifs}\left[\int_{-\infty}^{\infty} g_2(t-s)e^{-2\pi i(t-s)f}\, dt\right] ds$$

$$= \int_{-\infty}^{\infty} g_1(s)e^{-2\pi ifs}G_2(f)\, ds$$

$$= G_1(f)G_2(f)$$

Thus we have the important result that *the Fourier transform of a convolution of two functions is the product of their Fourier transforms.*

The preceding result has been shown in the time domain, but, by symmetry of the transforms, it applies equally in the frequency domain.

Occasionally the convolution of a function with respect to itself is of some interest. We have

$$h(t) = \int_{-\infty}^{\infty} g(s)g(t-s)\, ds$$

Applying the Fourier transform twice gives

$$h(t) = \int_{-\infty}^{\infty} G^2(f)e^{-2\pi ift}\, df$$

and for $t = 0$ we have

$$h(0) = \int_{-\infty}^{\infty} g(s)g(-s)\, ds = \int_{-\infty}^{\infty} G^2(f)\, df$$

Written in the form

$$\int_{-\infty}^{\infty} |g(s)|^2\, ds = \int_{-\infty}^{\infty} |G(f)|^2\, df$$

It is known as *Parseval's theorem.*

Exercises

8.7-1   Fill in the details for the second form of the convolution theorem.

8.7-2   Fill in the details for Parseval's theorem.

8.8   THE EFFECT OF A FINITE SAMPLE SIZE

Very often the time function $g(t)$ can be regarded as an infinitely long signal (in time). Of necessity, we must take a finite-length sample. Thus in astronomy we can observe a pulsar or a Cepheid variable for a finite length of time only. What does this limitation do to the original signal? We can regard the limitation as being equivalent to multiplication in time by the rectangular pulse of unit height $2Tp(t)$ (which is a multiplying window), where

$$p(t) = \begin{cases} \dfrac{1}{2T}, & |t| < T \\ 0, & |t| > T \end{cases}$$

This is also known as the "box car function," "gate function," and the "cookie cutter function." Thus what we see is not $g(t)$ but rather $g_1(t)$, defined by (see Section 8.5)

$$g_1(t) = 2Tg(t)p(t)$$

The Fourier transform is, by the convolution theorem,

$$G_1(f) = 2T \int_{-\infty}^{\infty} G(f_1) \frac{\sin 2\pi T(f_1 - f)}{2\pi T(f_1 - f)} \, df_1$$

where f_1 is the dummy variable of integration. So the transform of what we see is the convolution of the true transform $G(f_1)$ with the convolving window (f_1 is the variable)

$$2T \frac{\sin 2\pi T(f_1 - f)}{2\pi T(f_1 - f)}$$

In the situation in which the original signal is a single frequency (a spike or line in the spectrum), this line is convolved with the preceding sinusoidal function, and therefore we will see this same function, the Gibbs phenomenon in a slightly new form. It is the usual $(\sin f)/f$ function with the amplitude of the main lobe growing in height like $2T$ and the half-width to the nearest zero contracting like $1/2T$.

If we have many lines in the spectrum, instead of a single line, each will be smeared out by the same $(\sin f)/f$-type function, and in the limit of a continuous spectrum we will get the foregoing convolution integral. In an optical analogy used earlier, this smearing out corresponds to the idea of resolving power—the longer the time signal is observed, the better we can resolve (separate) adjacent spectral lines. Although intuitively evident that something similiar should occur, we have exhibited the actual dependence on T, where $2T$ is the length of the observation. Thus we can now understand what happens to a function when we chop out a piece of it to observe closely. The length of the sample we take limits what we can hope to discover from the sample.

If there are two lines, or other features, in the spectrum that are close to each other, then unless we have a long run of observations, the convolving window will not let us distinguish them from a single line. The closer the two lines are to each other, the longer the length of the observation must be if we are to resolve them. In a sense, the situation is similiar to two competing teams. The closer the teams are in ability, the longer we must observe them in order to decide which is better.

This result, that "chopping out" a segment from a continuous function amounts to convolving the transform of the function with

$$2T\frac{\sin 2\pi T(f_1 - f)}{2\pi T(f_1 - f)}$$

should be compared with the earlier "chopping out" of $2N + 1$ terms from a Fourier series in Section 5.5, where we obtained the corresponding convolving function

$$h(t) = \frac{\sin (N + 1/2)t}{\sin (t/2)}$$

and the modified window

$$h(t) = \frac{\sin Nt \cos t/2}{\sin t/2}$$

The comparison (allowing for the trivial fact that one is in the frequency variable and one is in the time variable) shows the relationship of the continuous model of the original signal and the discrete sampled model used in the actual data processing.

8.9   WINDOWS

Now that we have the formal apparatus of the Fourier integral, the subject of windows is a little clearer. In the domain of continuous functions (and not just transfer functions), the convolution theorem shows what happens to the continuous functions that we tend to hold in our minds; it is the discrete-continuous convolution theorems that tell us what occurs in the actual case being dealt with—that is, the discrete samples of the function. For both cases, windowing in one domain amounts to multiplication by the transform of the window, discrete or continuous, in the other domain. And it is inevitable; we can do nothing about it. The chopping out of a sample from an infinitely long function is windowing. Window theory tells us what to expect, as well as what to do rather than use a rectangular window if we want to reduce the effects of the discontinuities.

The use of a shaped window like the raised cosine of Section 5.9 has a similar result in the theory of Fourier integrals. The Fourier transform of

$$h(t) = \begin{cases} \dfrac{1 + \cos \pi t/T}{2}, & |t| < T \\ 0, & |t| > T \end{cases}$$

is

$$H(f) = \frac{1}{2T} \int_{-T}^{T} \frac{1 + \cos \pi t/T}{2} e^{-2\pi i f t} \, dt$$

$$= \frac{1}{4T} \int_{-T}^{T} [e^{\pi i t/T} + 2 + e^{-\pi i t/T}] e^{-2\pi i f t} \, dt$$

$$= \frac{1}{4} \frac{\sin 2\pi T(f - 1/2T)}{2\pi T(f - 1/2T)} + \frac{1}{2} \frac{\sin 2\pi f T}{2\pi f T} + \frac{1}{4} \frac{\sin 2\pi T(f + 1/2T)}{2\pi T(f + 1/2T)}$$

$$= \frac{1}{4} \frac{\sin 2\pi (f T - 1/2)}{2\pi (f T - 1/2)} + \frac{1}{2} \frac{\sin 2\pi f T}{2\pi f T} + \frac{1}{4} \frac{\sin 2\pi (f T + 1/2)}{2\pi (f T + 1/2)}$$

which can be rearranged by expanding the sines of the sums in both end numerators ($\sin \pi = 0$, $\cos \pi = -1$)

$$H(f) = \frac{1}{4} \frac{\sin 2\pi f T}{\pi} \left[-\frac{1}{2f T - 1} + \frac{2}{2f T} - \frac{1}{2f T + 1} \right]$$

Additional algebra results in the form analogous to the one in Section 5.9

$$H(f) = \frac{1}{2} \frac{\sin 2\pi f T}{2\pi f T} \left[\frac{1}{1 - (2fT)^2} \right]$$

The Hamming window in the continuous case may similarly be derived by following the method of Section 5.10. Alternately, we may note that the operations of smoothing and sampling may be interchanged (except for end effects at most) and deduce that the corresponding continuous case will have the appropriate form. We can see by Fig. 5.10-1 that the values of the weights in the limiting case approach the continuous case values, 0.23, 0.54, 0.23, as they should.

9

Kaiser Windows
and Optimization

9.1 WINDOWS

The purpose of this chapter is to improve on the earlier nonrecursive filter design methods of Chapter 7. Before doing so, however, let us review where we are and where we have been in order to clarify what we are doing. We began with digital filters, which are linear combinations of the data. Behind the design of filters is the data that is to be processed by them. We are still looking at nonrecursive filters, which use only the data and not the values that have been computed from the data. Furthermore, we are looking at symmetric (even) and skew-symmetric (odd) filters and not at the general case of arbitrary coefficients. If the original signal is

$$x(t)$$

we sample and quantize this function at a uniform spacing of unit time. Thus we observe the equally spaced data

$$x_n$$

and may have caused aliasing by the act of sampling.

The x_n are being processed by a time-invariant filter of the form (the coefficients c_k do not depend on n)

$$y_n = \sum_{k=-\infty}^{k=\infty} c_k x_{n-k}$$

In the general case, there can be infinite number of terms in the Fourier expansion, but, in practice, the sum runs between finite limits

$$y_n = \sum_{k=-N}^{k=N} c_k x_{n-k}$$

Since the system is linear, it follows that an eigenfunction

$$x(t) = Ae^{i\omega t} = Ae^{2\pi i f t}$$

will produce the output from the filter

$$y_n = \tilde{H}(f)Ae^{2\pi i f n}$$

where $\tilde{H}(f)$ depends on the c_k (and, of course, on f). Thus, as noted, the transfer function $\tilde{H}(f)$ is the corresponding eigenvalue. And a sum of such eigenfunctions

$$x_n = \sum_{k=1}^{K} A_K e^{2\pi i f_k n}$$

will give the corresponding sum at the output

$$y_n = \sum_{k=1}^{K} A_K \tilde{H}(f_k)e^{2\pi i f_k n}$$

Now that we have the theory of the Fourier integral, we know that any reasonable function $g(t)$ has a representation (a decomposition into frequencies) of the form

$$g(t) = \int_{-\infty}^{\infty} G(f)e^{2\pi i f t}\, df$$

The $G(f)$ corresponds to the A in the foregoing eigenfunction discussion, so that the output from the filter is

$$g_1(t) = \int_{-\infty}^{\infty} \tilde{H}(f)G(f)e^{2\pi i f t}\, df$$

For most nonrecuisive filters of the smoothing, stop, pass, band, and interpolation forms, we have

$$c_k = c_{-k}$$

And for them the transfer function

$$\tilde{H}(f) = \sum_{k=-N}^{N} c_k e^{2\pi i f k}$$

can also be written

$$\tilde{H}(f) = c_0 + 2 \sum_{k=1}^{N} c_k \cos 2\pi f k$$

The period of the transfer function is that of the Nyquist interval, and it arises from the sampling process with the corresponding aliasing. Thus, for practical purposes, the transfer function has meaning only in the interval

$$-\frac{1}{2} \le f \le \frac{1}{2}$$

9.2 REVIEW OF GIBBS PHENOMENON AND THE LANCZOS WINDOW

Consider once more the lowpass filter and its transfer function. In Section 6.3 we found its Fourier series

$$\tilde{H}(f) = 2f_s + 2 \sum_{k=1}^{k=\infty} \left[\frac{1}{\pi k} \sin 2\pi k f_s \right] \cos 2\pi k f$$

In practice, this series must be truncated to a finite number of terms. The process is equivalent to multiplying the terms of the Fourier series in its complex form by the corresponding terms of the series ($2N + 1$ nonzero terms)

$$0, 0, 1, 1, \ldots, 1, 0, 0,$$

In Section 5.7 we showed that this procedure is equivalent to convolving the transfer function $\tilde{H}(f)$ with (notice that we change from t to $2\pi f_1$ as the independent variable)

$$W(f_1) = \frac{\sin \pi (2N + 1)f_1}{\sin \pi f_1}$$

Thus the transfer function represented by the truncated Fourier series is the *convolution* of the ideal transfer function with this ratio of two sines,

$$\tilde{H}_1(f) = \int_{-\infty}^{\infty} \tilde{H}(f_1) W(f - f_1) \, df_1$$

We picture this convolution of the transfer function $\tilde{H}(f_1)$ with the ratio of sinusoids $W(f_1)$ (which appears as a gradually dying out sinusoid starting at a middle arch that is twice as wide as the other intervals between zeros), and we watch as the tail approaches the edge at the frequency, $-f_s$ (Fig. 9.2-1), of the lowpass filter being designed. We see that it is the ratio of the

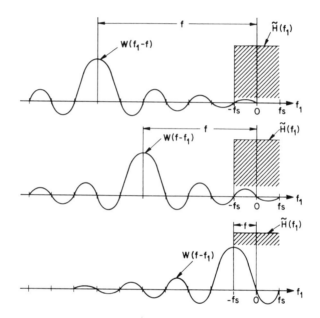

FIGURE 9.2-1 GIBBS PHENOMENON

two sinusoids that is integrated by the rectangular shape. Thus we must think about the integral of the window times the sinusoid ratio. As the convolution moves across the interval, more and more of the sinusoid ratio comes into the rectangular pulse; and when the main arch enters, a great rise in the value of the integral occurs. Finally, as the convolution continues, we see the dying out of the ripples in the integral due to the first edge. There is, of course, also the second edge at f_s of the filter to be considered. This figure displays the Gibbs phenomenon exactly; we see it in a new light. When we remember that both the ratio of the sinusoids and the original filter shape are periodic, we have a complete and accurate understanding of how the Gibbs phenomenon occurs as a result of truncation of the original series.

Recall that in Section 5.3 we looked at Lanczos' suggestion of how to reduce the Gibbs effect due to truncation. He suggested convolving the original rectangular function with a square window whose width is adjusted to the length of the ripple in the Gibbs wiggles. This process, we found, was equivalent to multiplying the coefficients of the Fourier series by the sigma factors, and it gave a reduction of almost a factor of ten in the ripple size.

A repetition of the Lanczos smoothing was seen to be equivalent to a triangular window of twice the width, and it produced the square of the sigma factor, thereby making the effect of the higher frequencies even less but also making the transition band twice as wide. We also looked at von Hann and Hamming windows.

This discussion suggests that we look further at what shape of window to use when truncating the Fourier series.

9.3   THE KAISER WINDOW: I_0-sinh WINDOW

What do we want for a window? We would like both the window and its transform to be narrow. But this situation is impossible in reality. How shall we compromise? If we were in the domain of continuous variable, then the *prolate spheroidal functions* are known to have the property that in a certain sense both are limited as much as possible. However, we are in the discrete domain of Fourier coefficients and have seen that there is a difference in the two cases. Reasoning that the difference between the two is not great in practice and that only a good approximation is needed to do the job, J. F. Kaiser suggested that we use as weights on the Fourier coefficients

$$w(k) = \begin{cases} \dfrac{I_0[\alpha\sqrt{1-(k/N)^2}]}{I_0(\alpha)}, & |k| \leq N \\ 0, & |k| > N \end{cases}$$

in place of the sigma factors of Lanczos, where

$$I_0(x) = 1 + \sum_{n=1}^{\infty} \left[\frac{(x/2)^n}{n!}\right]^2$$

Notice how Kaiser's weights resemble the Hamming raised cosine on a platform, since at the ends of the nonzero terms we have the value $w(N) = 1/I_0(\alpha)$, whereas at the middle we have the value $w(0) = 1$. The parameter α thus determines the height of the platform.

Kaiser's weights contain two parameters—N, which is the half width of the window (where we keep $2N + 1$ complex Fourier coefficients), and α, which controls the "shape" of the window, in particular, how large the ripples will be.

It can be shown that the transform of the $w(k)$, now regarded as a continuous function, is the function

$$W(f) = \frac{2N \sinh \left[\alpha\sqrt{1 - (f/f_a)^2}\right]}{\alpha I_0(\alpha)\sqrt{1 - (f/f_a)^2}}$$

where

$$f_a = \frac{\alpha}{N}$$

Notice that it is a sinh function when $f < f_a$, but for $f > f_a$ it becomes

$$\frac{2N \sin \left[\alpha\sqrt{(f/f_a)^2 - 1}\right]}{\alpha I_0(\alpha)\sqrt{(f/f_a)^2 - 1}}$$

and is sinusoidal with decaying ripples (due to the increasing size of the denominator) of amplitude approximately $1/f$; see Fig. 9.3-1.

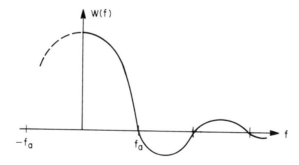

FIGURE 9.3-1 KAISER WINDOW

In the transform domain, $W(f)$ is convolved with the rectangular shape of the ideal filter; thus it is the convolution of the transform $W(f)$ that makes the ripples in the frequency domain. The maximum allowed ripple will be called δ in both the stop and pass bands.

This convolution was computed numerically by Kaiser. Table 9.3-1 gives a quantity α related to the maximum overshoot as a function of attenuation A (in decibels), where we will explain the symbols in detail in the following example. We note for future reference that for $\alpha = 0$ we have the pure case of Gibbs with no window shaping at all, since $I_0(0) = 1$. The $\alpha = 5.4414$ corresponds to the Hamming window, white $\alpha = 8.885$ corresponds to the

TABLE 9.3-1   Performance of I_0-sinh window, for equally spaced values of the attenuation

A (dB)	α	D
25	1.333	1.187
30	2.117	1.536
35	2.783	1.884
40	3.395	2.232
45	3.975	2.580
50	4.551	2.928
55	5.102	3.276
60	5.653	3.625
65	6.204	3.973
70	6.755	4.321
75	7.306	4.669
80	7.857	5.017
85	8.408	5.366
90	8.959	5.714
95	9.501	6.062
100	10.061	6.410

Blackman window [BT, p.98]. The third column of the table gives the window width D.

Before becoming enmeshed in the details of how the formulas occur, we will first go through the steps of the design of a filter.

We begin by making a sketch of the ideal filter (Fig. 9.3-2) that we would like to have. This figure includes the acceptable width ΔF of the transition band (bands) and δ, the ripple half-amplitude that we can tolerate.

Second, we calculate the attenuation A in dB

$$A = -20 \log_{10} \delta$$

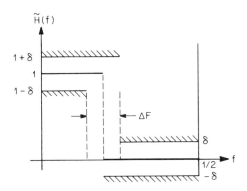

FIGURE 9.3-2   IDEAL FILTER

From the attenuation A we find the α to shape the window [because the tails of $W(f)$ produce the ripples in the convolution]. To do so, we use the empirical formula that Kaiser derived by curve fitting Table 9.3-1.

$$\alpha = \begin{cases} 0.1102(A - 8.7), & A > 50 \\ 0.5842(A - 21)^{0.4} + 0.07886(A - 21), & 21 < A < 50 \\ 0 & A \leq 21 \end{cases}$$

A sketch of this function appears in Fig. 9.3-3. In this figure we see that the Gibbs 8.9% overshoot (21-dB attenuation) corresponds to the value $A = 21$.

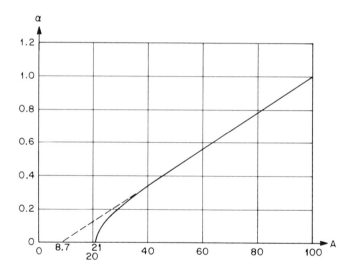

FIGURE 9.3-3 α VS ATTENUATION

Finally, the number of terms to keep, $(2N + 1)$, is given by the formula for N,

$$N = \frac{A - 7.95}{28.72\,\Delta F}$$

This formula shows that N is inversely proportional to the width of the transition band ΔF. The attenuation A depends on the log of the ripple height δ, and thus it is much easier to reduce the ripple size than it is to narrow the transition band, as judged by the number of terms in the final filter. In this way, we find $w(k)$ and $W(f)$ for the windowing of the original Fourier series approximation.

9.4   DERIVATION OF THE KAISER FORMULAS

We begin the derivation of the formulas by recalling that the span of a nonrecursive filter of $2N + 1$ terms is $2N$ intervals and that we used the normalized frequency so that the Nyquist interval reaches from $-\frac{1}{2}$ to $\frac{1}{2}$. The normalized transition width is ΔF.

Kaiser noted that, for a fixed δ, the product of the two widths—the window span and the transition width—was approximately a constant; it was defined as the D factor

$$(2N)\,(\Delta F) = D$$

It is now necessary to find a formula for N that gives the number of terms in the digital filter.

By numerical integration of the sinh function, Kaiser found Table 9.3-1. The overshoot gives the attenuation, and the first crossing into this admissible band gives $\Delta F/2$ and hence D. From this table Kaiser derived the empirical formulas for α and D, α as above and D by

$$D = \begin{cases} \dfrac{A - 7.95}{14.36}, & A > 21 \\ 0.9222, & A < 21 \end{cases}$$

Solving for N above, we get

$$N = \frac{D}{2\,\Delta F} = \frac{A - 7.95}{14.36(2\,\Delta F)}$$

for the number of terms.

9.5   DESIGN OF A BANDPASS FILTER

We have approached the design problem as if it were a low- (or high) pass filter, but the same methods apply to bandpass filters with one warning, which will be mentioned later.

Suppose, as in Fig. 9.5-1, that we want to design a bandpass filter. As usual, we expand the ideal transfer function in the Fourier series

$$\tilde{H}(f) = c_0 + 2 \sum_{k=1}^{N} c_k \cos 2\pi k f$$

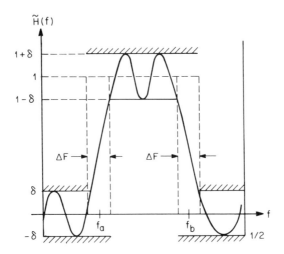

FIGURE 9.5-1 BAND PASS FILTER

where the c_k are the coefficients in the digital filter being designed. We get for the c_k

$$c_0 = 2(f_b - f_a)$$

$$c_k = \frac{1}{k\pi}[\sin 2\pi k f_b - \sin 2\pi k f_a]$$

For a lowpass filter, $f_a = 0$; and for a highpass filter, $f_b = \frac{1}{2}$.

From our δ we calculate the attenuation A, the window shape α, and, finally, the number of terms N (equivalent in the filter to $2N + 1$ coefficients). In the formula we round up N if necessary. Then we evaluate $w(k)$ at the N points, with $w(0) = 1$, $w(N) = 1/I_0(\alpha)$, and $w(k) = w(-k)$, and use the products $c_k w(k)$ as the final filter coefficients.

Having done so, we check the formula by evaluating the Fourier series of the transfer function. We will find that sometimes the transfer function has a δ that is too large. To see why a too-large δ can occur in a bandpass filter, recall that the ripples came from the tails of the transform, the sinh (or sin) formula, as they are convolved into the step of the ideal filter. Since there are two steps in each band (positive and negative frequency), four sets of ripples are created by the tails while the window is being convolved into the ideal filter characteristic. In the worst case, these ripples can combine and make the total ripple at most twice as large as it should be, 6 dB larger than we designed for. Clearly, this effect depends on the spacing of the two vertical sides of the bandpass filter, the spacing between the positive and negative frequency bands, and the spacing of the ripples in the window function, (the N we use). If this error cannot be tolerated, we can redesign the filter, starting with a smaller δ to allow for this effect.

9.6   THE SAME DIFFERENTIATOR AGAIN

In Section 6.4 we designed a differentiator that cut off ideally at $f = 0.2$ to remove noise in the upper 60% of the frequency band. Let us see how well the Kaiser window handles this same problem.

For comparison purposes, we will use the number of coefficients rather than the transition band width. (Solving for the transition width in terms of N is not difficult.)

Figures 9.6-1 and 9.6-2 show the transfer functions in cases 30 dB and 50 dB for $N = 5, 7, 10$ terms. The higher flatness at the upper end of the Nyquist interval is purchased at the price of a wider transition band. This factor is an asset if the noise is confined to the upper part of the interval, but if it is spread across the whole band, matters are rather different. In Fig. 9.6-3 we see another comparison of filters for $N = 5$. Finally, in Fig. 9.6-4 we have another comparison.

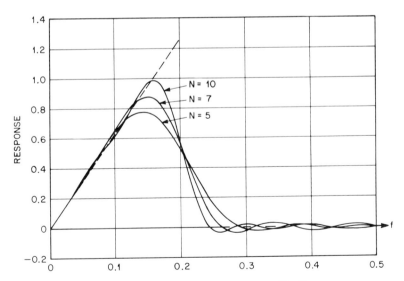

FIGURE 9.6-1 FREQUENCY RESPONSE OF A 30 dB DIFFERENTIATOR-SMOOTHING DIGITAL FILTER

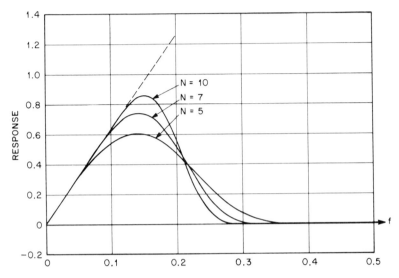

FIGURE 9.6-2 FREQUENCY RESPONSE OF A 50 dB DIFFERENTIATOR-SMOOTHING DIGITAL FILTER

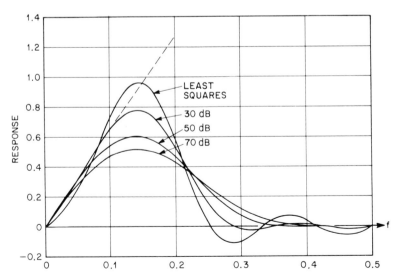

FIGURE 9.6-3 FREQUENCY RESPONSE OF A 30 dB DIFFERENTIATOR-SMOOTHING DIGITAL FILTER

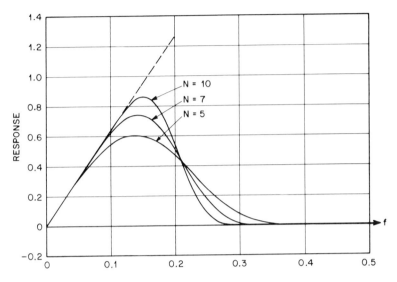

FIGURE 9.6-4 FREQUENCY RESPONSE OF A 50 dB DIFFERENTIATOR-
SMOOTHING DIGITAL FILTER

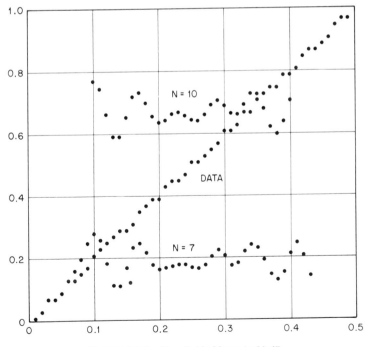

FIGURE 9.6-5 $N = 7, 10$; NOISE 1; 30 dB

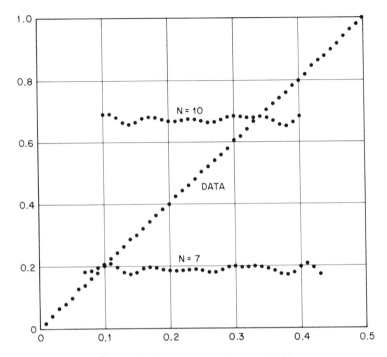

FIGURE 9.6-6 $N = 7, 10$; Noise 2; 30 dB

To test the filter on some data, we will use the same noise as in Section 6.4 so that the results can be compared easily. See Figs. 9.6-5 and 9.6-6.

9.7 A PARTICULAR CASE OF DIFFERENTIATION

This example is taken from nuclear physics and is given to provide some of the flavor of a real problem in filter design. The original data consisted of counts of nuclear events classified according to the energy of the particles. Thus we are given a table of how many particles have energies in each of the energy intervals (equally spaced). We see immediately that such data must be "noisy," since from experiment to experiment we would obtain rather different counts of the events in the various intervals. Furthermore, we can expect that the experiment was run for as long as practical but not as long as one could wish. Moreover, in practice, the square roots of the counts should be used, for they will have equal variance in the various intervals. However, we will not confuse the reader with this detail.

When in addition to this basic noise of the underlying phenomenon we observe that the physically meaningful quantity is the derivative of the nuclear counts, then we see that we have a very noisy situation that requires care in the processing of the data.

In order to measure where the signal and noise can be separated, two steps were taken. First, a spectrum of the raw data was computed (see Section 10.7). It showed that the spectrum was essentially flat past about the first 5% of the frequency interval. Second, a theoretical shape of the data was tried and again analyzed by the spectrum program. It, too, showed about the same breakpoint between the message and the noise. Why do we feel that a flat spectrum implies noise? Simply, if we have a noise spectrum falling off rather slowly, then the aliasing that arises from the size of the intervals (the folding back and forth of the spectrum) will usually tend to be flat across the whole spectrum, while the message in a properly designed experiment will not be aliased.

A differentiator filter was therefore designed that rose linearly from approximately $f = -1/40$ to $1/40$ in the frequency domain and was zero elsewhere. The Fourier expansion was naturally in sines with a spare factor i [which is necessary, since the derivative of $\exp(2\pi i f t)$ has a factor i]. This expansion, which varied, depending on the exact placement of the edge (so we could adjust this cutoff point), then served as the basis for the design. The Kaiser window was used to get the final filter design.

The nuclear counts were grouped into equal-sized intervals, and it was the total in each interval that was recorded. This procedure is not the same as recording samples of the function. However, if we imagine that the Lanczos window is convolved across the original function and that the result is sampled, we find that we have the number of counts in the interval. Thus the grouping of the data is equivalent to a preliminary pass of a Lanczos window and should be allowed for in the final interpretation.

As judged by the physicist, the results were satisfactory (of course, better data was desired, but the filter design caused no complaints). There is some risk that, by adjusting the cutoff edge on a single run of data, we will be fooled as to the accuracy; however, by applying it to several runs taken under different conditions, the effect of moving the edge was understood.

This example illustrates a number of important factors. First, the data need not be a time series but can be a series of equally spaced data in some other variable. Secondly, using the preceding methods of filter design, almost all the noise can be successfully eliminated from the interval where it does not overlap the signal. Thus filters apply not only to time series but to other forms of equally spaced data as well.

9.8   OPTIMIZING A DESIGN

Up to now we have adopted simple design criteria, the closeness (in a somewhat unspecified sense) of the actual transfer function to the ideal square-cornered transfer function. Both the raised cosine (with or without the platform) and the Kaiser windows are means of changing the Fourier series (least-squares) fit of the original design to reduce ripples (but, of course, they increase the error in a least-squares sense). In practice, there are numerous ways of specifying the properties desired for a particular filter.

To meet the different criteria, many design methods exist and more are easily invented. But such special methods are beyond the scope of this book. There is, fortunately, a very elementary design method that meets many such needs:

1. Decide on the kind of criterion that you want to apply and express it as a positive number with the reasonable order relationship that worse designs have larger numbers.
2. Find a first approximation to the filter that you want.
3. Optimize the design by varying the adjustable parameters of the filter, using (1) as a guide.

This elementary design method has a number of possible disadvantages. First, a great deal of machine time may be needed. Second, the optimization comes down to a *local* minimum, and it is possible that there is a much better design that can be reached from some other starting point. Finally, you have little or no idea of what similar filters with a slightly different number of terms will do unless you repeat the whole process; thus you are blindly searching for an optimum with almost no theory to guide you. This lack of elegance makes the experts tend to avoid the method; but for the beginner and the occasional designer it has the great advantage of wide applicability and, in addition, it requires comparatively little special knowledge.

In real life the difficulties are generally the choice of the criteria and the first trial transfer function to use and not the amount of machine time. Kaiser's design method should help guide the beginner in an actual situation toward making these choices, but, in practice, the choice of the proper criterion is outside the realm of an elementary course.

The slowness of the process of finding the optimum is less important. A minimum that is difficult to locate suggests a "delicate" design that probably should be avoided for practical reasons. Each year computer use becomes less and less expensive, and so its cost should not be overemphasized. However, the lack of surrounding theory, and the risk of finding a local optimum that is far from the best, are serious matters that should be considered in each application.

9.9 A CRUDE METHOD OF OPTIMIZING

Suppose that we are given both an evaluation function E that measures how well the criterion is met (E gives nonnegative values and decreasing E means a better design) and a first approximation to the design with the N parameters $p_1, p_2, \ldots, p_N$ (usually the coefficients c_k of the filter).

We set up three arrays in a computer; first, the array of N initially guessed coefficients (parameters) p_k from the approximate design; second, a corresponding array of N estimates of how much we think the coefficients might be in error Δp_k (these guesses will rapidly be eliminated and replaced with more accurate ones generated by the optimization routine so that bad guesses merely lead to extra computations); and, third, an array of the N improvements due to a change in the parameters (the third array is later filled in).

The first step in the computation is to evaluate the function E for the given coefficients p_k. We put $1/N$ of this amount into each place in the third array, where N is the number of parameters p_k.

The next step is to change the first parameter p_1 by the amount Δp_1, which is the entry in the uncertainty array corresponding to the first parameter, and again evaluate the function E. If the difference in the two E values is negative, we have found an improvement. Therefore we take another step of the same size in the same direction along the p_1 axis. We continue until we find an E value that is $\geq$ the previous one (note the $\geq$). We then have three consecutive values of the parameter p_1, which we will label p_1^-, p_1^0, p_1^+. If the first change in E was an increase or the same value, we interchange the starting and first trial values, along with the sign of Δp_1, and continue going in this direction, searching as before for a relative minimum of E as a function of p_1. When we find it, we are in the same position as before. If we discover that the value of E rises when we move in either direction, we have a situation similar to the preceding one of three values with the smaller value in the middle.

Through the three consecutive values we pass a parabola

$$E(p_1) = A + B(p_1 - p_1^0) + C(p_1 - p_1^0)^2$$

where p_1^0 is the middle value of the three values of the coordinate of the parameter p_1. The minimum of the parabola occurs at the value where

$$\frac{dE(p_1)}{dp_1} = 0 = B + 2C(p_1 - p_1^0)$$

Thus the minimum occurs at

$$p_1 = p_1^0 - \frac{B}{2C}$$

From the three given values it is easy to see that the coefficients of the parabola are given by

$$B = \frac{E(p_1^+) - E(p_1^-)}{2d}$$

$$C = \frac{E(p_1^+) - 2E(p_1^0) + E(p_0^-)}{2d^2}$$

where (the step spacing)

$$d = p_1^0 - p_1^- = p_1^+ - p_1^0$$

From our method of search we avoided having $C = 0$, and so we are not dividing by 0 (unless all three values of E are the same!). We now have a new p_1.

Having temporarily optimized the first parameter, we need to refill the corresponding values in the three arrays. Obviously the minimum of the parabola is taken as the new entry p_1 in the first array. We could pick one-half the change from the initial to the final estimates of p_1 as the new uncertainty, Δp_1, but we must avoid having the new uncertainty be zero, for if it were, then no further search would occur. Thus we take the larger of one-half of the change and one-half of the initial uncertainty as our new uncertainty. Finally, we select the change in the function E from its initial value to the value at the new point as the corresponding entry in the third array. Again we need to worry about a zero entry, since, as we shall see, it would be, impossible to resume examining improvements in the parameter if it were zero. To avoid this trouble we again adopt a selection of the larger of one-half of the original change in E and the current amount of the improvement in E.

With the first parameter processed, we repeat the procedure for the second, the third, and so on, through the entire set of parameters. We end up with three new arrays: the first has the new estimates for the parameters; the second has the uncertainties that reflect how much we moved from our initial guess to the new one; and the third array has an indication of where we achieved large and where we achieved small improvements. So once we have gone through the array, we have available the computer-produced estimates of the parameter values, the uncertainties, and where to expect the most improvement.

From this point on we scan the improvement array for the largest value and work on the corresponding parameter. As a result, at every stage of the computation we tend to have about the same order of magnitude in all the improvements in the third array. The corresponding step sizes that produce the improvements indicate the sensitivity of the fit in the parameters. When the improvements become small per computation cycle, we cannot expect that E will get less than the minimum allowed for the model that we are testing, and we stop further computation. Gradually increasing the values in the improvement column is worthwhile in order that a chance small value will not prevent us from looking there later.

Notice that this algorithm has a number of invariances. It rapidly eliminates the starting values, both the positions and the uncertainties, and also omits the chance order in which we selected the parameters of the problem. At the end of the computation each parameter change has about the same amount of effect in improving the fit as any other parameter.

This simple optimizer illustrates several important points. First, it is the kind of an optimizing process that the beginner is apt to believe is good. And it is good for simple problems.

Second, it shows how a simple optimizer can fail. Suppose, for instance (as we will be doing in Chapter 12), that we are trying to reduce the maximum ripple of a polynomial in some interval. It is probable that one parameter, say p_1, will tend to reduce one peak (or valley) of the polynomial while increasing the size of one or more other extremes. Thus changes in p_1 will work *until* some second extreme comes up to the same value as the first one. There will be a sharp "corner" to the surface at this point. It is also likely that some other parameter, say p_2, will tend to reduce the second extreme while at the same time increasing the first. Thus changes in neither parameter can by themselves improve matters once the two extremes are of the same size: only a combination of simultaneous changes in both will further reduce the maximum extreme of the polynomial in the given interval. And, of course, the more equal-sized extremes there are, the more parameters must simultaneously be changed if further improvements are to be found. The simple, one-parameter-at-a-time optimizer is clearly inadequate.

The third point is that powerful optimizers in the local library tend to be "powerful" in one sense or more. They may try to reach the optimum rapidly, but this fact only saves machine time and is not apt to be important to the occasional digital filter designer. They may be powerful in the sense that they will find the minimum for a wide range of surfaces, including situations where the valley being searched is both steep and curved, so that continually different directions of search must be used. In addition, the power may refer to the ability to pinpoint the minimum. This feature is often illusory. If the minimum is sharp and easy to locate, the performance of the filter will change greatly for small changes in the parameter values (perhaps a roundoff to a shorter length number), whereas if it is a broad minimum, then, even though the place where the true minimum occurs is not well determined, this fact is unimportant, since almost any nearby point will give almost equally good results.

Finally, the example selected shows that the minimizing surface being searched can have "corners" for changes in many of the parameters. Therefore if the local library optimizing routine is based on a smooth surface and fits the surface with quadratic terms in all the parameters, it is possible that the optimizer will behave peculiarly.

10

The Finite
Fourier Series

10.1 INTRODUCTION

We have at times discussed the continuous functions of frequency as if it were possible to handle continuous functions in a computer, but, of course, we are forced to use samples of such functions. Fortunately, the sines and cosines are orthogonal over *both* the continuous interval and sets of equally spaced points. Thus we can perform analogous operations in the discrete versions of the functions used in a computer, and aliasing must be considered once more.

The problems faced in this chapter are (a) to prove the orthogonality of the sines and cosines over a set of equally spaced points, (b) to relate the continuous and discrete expansions (via aliasing), and (c) to give a sketch of the ideas behind the fast Fourier transform method of computing the coefficients of the discrete expansion (most installations have their favorite versions in the library). The fast Fourier transform is simply a means of computing the Fourier transform in essentially $N \log N$ arithmetic operations instead of—what might at first seem necessary—N^2 operations. This difference is of fundamental importance to many applications, since for large N it can mean a reduction by factors of hundreds, or thousands, or even more in machine time used, as well as a significant reduction in roundoff errors in the results.

10.2   ORTHOGONALITY

The discussion will be restricted to an even number of points, with the corresponding case of an odd number of points left to the reader in case he ever faces this likely situation; there are only trivial differences between the two cases. Let the $2N$ points we have be

$$x = 0, \frac{L}{2N}, \frac{2L}{2N}, \ldots, \frac{(2N-1)L}{2N}$$

Using p as the index, these numbers can be written

$$x_p = \frac{Lp}{2N} \qquad (p = 0, 1, 2, \ldots, 2N-1)$$

It is easiest to begin with the orthogonality of the complex exponentials, and to do so, we begin with the very simple geometric progression (using the $2N$ points x_p)

$$\sum_{p=0}^{2N-1} e^{(2\pi i q/L)x_p} = \sum_{p=0}^{2N-1} e^{\pi i pq/N} \qquad \text{(for integer } q\text{)}$$

This geometric progression has the ratio

$$r = e^{\pi i q/N} \qquad r^{2N} = e^{2\pi i q} = 1$$

and the sum is

$$\begin{cases} \dfrac{1 - r^{2N}}{1 - r} = 0, & r \neq 1 \\ 2N, & r = 1 \end{cases}$$

The situation of $r = 1$ arises only when $q = 0, \pm 2N, \pm 4N, \cdots$. These values of q, as we shall show, lead to the aliasing that we know must occur as a result of the equally spaced sampling of the function.

From this simple summation we move to the set of functions (k is an integer)

$$e^{(2\pi i/L)kx_p}$$

and prove that they are orthogonal. "Orthogonal" means that the summation over the set of sample points of the kth function of the set times the complex conjugate of the mth is zero, except when the same function, or an equivalent aliased function, is used for both. Thus we need to prove

$$\sum_{p=0}^{2N-1} e^{(2\pi i/L)kx_p} e^{-(2\pi i/L)mx_p} = \begin{cases} 0, & |k-m| \neq 0, 2N, 4N, \ldots \\ 2N, & |k-m| = 0, 2N, 4N, \ldots \end{cases}$$

which follows immediately by writing the product of the exponentials as a single exponential and applying the previous result, with $q = k - m$.

To get to the "real functions" of sine and cosine, we use the Euler identity

$$e^{ix} = \cos x + i \sin x$$

Writing out the sum of the product of two trigonometric functions, we obtain both the sum and difference of the exponents if the two functions are both sine or both cosine. We have for the cosines (notice the aliasing)

$$\sum_{p=0}^{2N-1} \cos \frac{\pi kp}{N} \cos \frac{\pi mp}{N} = \begin{cases} 0, & |k \pm m| \neq 0, 2N, 4N, \ldots \\ N, & |k+m| \text{ or } |k-m| = 0, 2N, 4N, \ldots \\ 2N, & |k \pm m| \text{ both } = 0, 2N, 4N, \ldots \end{cases}$$

Similarly, for the sines we have

$$\sum_{p=0}^{2N-1} \sin \frac{\pi kp}{N} \sin \frac{\pi mp}{N} = \begin{cases} 0, & |k \pm m| \neq 0, 2N, 4N, \ldots \\ N, & |k-m| = 0, 2N, 4N, \ldots \\ -N, & |k+m| = 0, 2N, 4N, \ldots \\ 0, & k = m = 0 \end{cases}$$

The case of a sine times a cosine is easily shown to be always zero.

For the restricted set of functions

$$1, \cos \frac{2\pi x}{L}, \cos \frac{4\pi x}{L}, \ldots, \cos \frac{2\pi(N-1)x}{L}, \cos \frac{2\pi N x}{L}$$

$$\sin \frac{2\pi x}{L}, \sin \frac{4\pi x}{L}, \ldots, \sin \frac{2\pi(N-1)x}{L}$$

only one of the conditions for a nonzero sum can apply at a time, and we have the orthogonality of the $2N$ Fourier functions over the $2N$ discrete equally spaced points x_p. This orthogonality is the same as we had over the continuous interval, with, of course, different normalizing factors. Thus we are led to the Fourier expansion of an arbitrary function $G(x)$ defined over the discrete set of $2N$ points x_p

$$G(x) = \frac{A_0}{2} + \sum_{k=1}^{N-1} \left[A_k \cos \frac{2\pi k x}{L} + B_k \sin \frac{2\pi k x}{L} \right] + \frac{A_N}{2} \cos \frac{2\pi N x}{L}$$

where we are using capital letters for the coefficients in the discrete expansion. Note that the first and last cosine terms have an extra $\frac{1}{2}$ factor. We have from the orthogonality

$$A_k = \frac{1}{N} \sum_{p=0}^{2N-1} G(x_p) \cos \frac{2\pi k x_p}{L} \qquad (k = 0, 1, \ldots, N)$$

$$B_k = \frac{1}{N} \sum_{p=0}^{2N-1} G(x_p) \sin \frac{2\pi k x_p}{L} \qquad (k = 1, 2, \ldots, N-1)$$

Frequently the function $G(x)$ is given at $2N + 1$ points, *including both ends of the* interval, and we still have (because of periodicity) only $2N$ intervals. In such cases, since we have assumed that the function is periodic, we usually take the true end value as the average of the two endpoints,

$$\frac{G(0) + G(L)}{2}$$

Of course, if the function is periodic, then both end values are the same, and this is simply one of the values. The effect of this averaging is to expand slightly the basic formulas for the coefficients of A_k without changing the B_k.

We see that the summation formula for A_0 in the discrete case corresponds to the trapezoid rule of integration. If we used the midpoint formula rather than the trapezoid rule for approximating the integral, above arguments would lead to a corresponding set of orthogonality relations over these midpoints of the intervals. For these points the highest frequency cosine term will be identically zero, and we need instead to include the corresponding sine term in the basic set of functions.

Exercises

10.2-1   Sum the geometric progression over $2N + 1$ points.

10.2-2   Prove the orthogonality of the sines and cosines over $2N + 1$ points.

10.2-3   Using the $2N$ points (midpoints of the $2N$ intervals)

$$x_p = \frac{L(1 + 2p)}{4N} \qquad (p = 0, 1, \ldots, 2N)$$

prove the orthogonality of $2N$ functions.

10.2-4   Using Exercise 10.2-3, derive the formulas for the coefficients of the expansion. Discuss the small differences in the two cases.

10.3 RELATIONSHIP BETWEEN THE DISCRETE AND CONTINUOUS EXPANSIONS

Since it is occasionally necessary to replace the continuous transfer functions of our theory with the discrete functions of computing practice, we investigate their mutual relationship. Suppose that the continuous expansion of $G(x)$ is given by

$$G(x) = \frac{a_0}{2} + \sum_{k=1}^{\infty} \left[a_k \cos \frac{2\pi x}{L} + b_k \sin \frac{2\pi x}{L} \right]$$

where we have used lowercase letters for the coefficients of the continuous function expansion and will continue to use uppercase letters for the discrete expansion. If we multiply both sides of this equation by $\cos 2\pi k x_p / L$ and sum, we obtain, due to the aliasing given in the previous section,

$$\sum_{p=0}^{2N-1} G(x_p) \cos \left(\frac{2\pi k x_p}{L} \right) = N A_k = N(a_k + a_{2N-k} + a_{2N+k} + \cdots)$$

from which we see that the coefficient calculated in the expansion A_k is given by

$$A_k = a_k + \sum_{m=1}^{\infty} (a_{2Nm-k} + a_{2Nm+k})$$

For each k this formula gives the calculated result A_k in terms of the true answers a_k. It shows the usual aliasing in a slightly new form and offers a different derivation of the aliasing effect. The aliasing is illustrated in Fig. 10.3-1. Similar calculations using the sine in place of the cosine lead to the

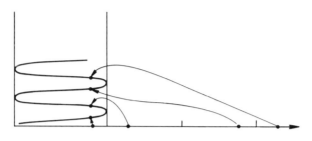

FIGURE 10.3-1   ALIASING

formula

$$B_k = b_k + \sum_{m=1}^{\infty} (-b_{2Nm-k} + b_{2Nm+k})$$

The constant term gives an especially important result, for it shows the relationship in the frequency domain between the integral of a periodic function and its trapezoid rule approximate calculation value A_0

$$A_0 = a_0 + 2 \sum_{m=1}^{\infty} a_{2Nm}$$

If the higher-indexed coefficients are small, then the error is small.

Thus we see the role played by the Nyquist folding frequency in a different light; we can act as if we had a continuous function, but the process of sampling will bring us into the Nyquist interval (or any other equivalent interval that we care to choose).

Exercise

10.3-1   Carry out all the details for the aliasing expressions for A_k and B_k.

10.4   THE FAST FOURIER TRANSFORM

The direct approach to the calculation of the Fourier coefficients of a discrete expansion appears at first to be of the order N^2 operations; there are $2N$ coefficients and $2N$ terms in each summation. A recent rediscovery and adequate presentation by Cooley and Tukey of a method for calculating the Fourier coefficients reduces the amount of computing to the order of N log

N. This amount can be significant in a large problem, over a factor of 100 in machine time saved—less than 1 % of the original time.

Some research is still being done on how to extract the last bit of savings of machine time, but this factor is of little interest in an introductory course. We therefore present the main idea and avoid becoming involved in details. As a rule, the local computer library has a fast Fourier transform program. It is important to emphasize, however, that within roundoff the results of the straightforward method and the fast Fourier transform method are exactly the same; they are simply alternate ways of computing the same thing. The fast Fourier transform, because it does less arithmetic, typically has less roundoff error in the results. For convenience, we will set $L = 1$ so that $x_p = p/2N$.

We assume that the number of sample points of the function, $2N$ (or $2N + 1$), can be factored into two integers greater than 1

$$2N = PQ$$

Thus the sample points are

$$x_m = \frac{m}{PQ}$$

The Fourier coefficients of the expansion of the function $G(x)$ are

$$A_k = A(k) = \frac{1}{PQ} \sum_{m=0}^{PQ-1} G(x_m)e^{-2\pi i k x_m}$$

We now do the decisive step and divide k by p in order to obtain the quotient k_1 and the remainder k_0. The next step is to divide m by Q and get the quotient m_1 and the remainder m_0. Both k and m are *uniquely* represented when we write the two summation variables as

$$k = k_0 + k_1 P$$
$$m = m_0 + m_1 Q$$

where, of course, we have the conditions

$$k_0 < P \quad \text{and} \quad k_1 < Q$$
$$m_0 < Q \quad \text{and} \quad m_1 < P$$

Using these representations for k and m, we then have for the summation (note that we have taken into account all the terms but have used the essential relation $e^{-2\pi i k_1 P_1} = 1$)

$$A(k_0 + k_1 P) = \frac{1}{PQ} \sum_{m_0=0}^{Q-1} \left[\sum_{m_1=0}^{P-1} G\left(\frac{m_0 + m_1 Q}{PQ}\right) e^{-2\pi i (k_0 + k_1 P)(m_0 + m_1 Q)/PQ} \right]$$

$$= \frac{1}{PQ} \sum_{m_0=0}^{Q-1} e^{-2\pi i k_0 m_0/PQ} e^{-2\pi i k_1 m_0/Q} \left[\sum_{m_1=0}^{P-1} G\left(\frac{m_0}{PQ} + \frac{m_1}{P}\right) e^{-2\pi i k_0 m_1/P} \right]$$

It is easy to recognize that the quantity in the square brackets is a Fourier expansion involving $1/Q$ of the samples of the function, phase shifted m_0/PQ. Furthermore, since $0 \le m_0 < Q$, there are Q such sums to be done. By labeling these sums as

$$\bar{A}(k_0, m_0)$$

we will have the formula

$$A(k_0 + k_1 P) = \frac{1}{PQ} \sum_{m_0=0}^{Q-1} \bar{A}(k_0, m_0) e^{-2\pi i [(k_0/PQ) + (k_1/Q)] m_0}$$

This is a second Fourier series expansion computation, and this time, since $0 \le k_0 < P$, there are P such sums to be evaluated.

A count of the number of arithmetic operations shows that it is proportional to

$$PQ(P + Q)$$

This amount of computation is in place of the original amount which was proportional to $(PQ)^2$. Evidently, for each factor P_i greater than 1 that we can find in the number $2N$, we can repeat this process. In the end we will arrive at, approximately,

$$P_1 P_2 P_3 \cdots P_k (P_1 + P_2 + \cdots + P_k)$$

arithmetic operations instead of the original $(2N)^2$ operations. Thus we see that when the number of sample points $2N$ is a power of 2, we will obtain the maximum reduction of computation, and the number of terms appearing in the sum inside the parentheses is the $\log (2N)$.

Experience shows that this simple trick covers most of the savings (in a multiplicative sense) and that additional refinements in the details will save relatively less. However, if the Fourier transform is to be done frequently, then further tricks can and should be used to reduce roundoff even if the computing time saved is not important. See references [Br] and [IEEE-1].

The fast Fourier tranform provides a powerful method of filter design. Given the original data x_n, we take the Fourier transform of the x_n. Next, we take the values, which are the Fourier expansion coefficients, and multiply them by their corresponding transfer function values. Since the transfer function is the curve of gain (or attenuation) of the various frequencies, this multiplication is equivalent to filtering. Then we need only apply the fast Fourier transform to these products to obtain the filtered data.

The entire process is as follows: (a) transform the data, (b) multiply the results by the transfer function, and (c) transform the products back! It is a very general design method for nonrecursive filters. The reader may, of course, be a little concerned over what happens to frequencies between the sample points of the transfer function that were used. Yet if the points are densely placed, there should be no problem (except near sudden jumps where Gibbslike wiggles can be expected). At a discontinuity of the transfer function, of course, the average of the two limits is used.

10.5   COSINE EXPANSIONS

Since the transfer functions are often even functions, that is

$$\tilde{H}(f) = \tilde{H}(-f) \qquad \left(|f| \leq \frac{1}{2} \right)$$

it is worth a special effort to look at the corresponding Fourier expansions.

All the coefficients of the sine terms are zero. This result is immediately obvious from the fact that the sine is an odd function of its argument and that the product of an even and an odd function is an odd function, so that the integration over a symmetric interval is zero. Thus only the cosine terms need be considered. Again, from the evenness, this time we see that the summations need be done over only half the range, and we will get

$$A_k = \frac{1}{2N} \left[G(0) + 2 \sum_{p=1}^{N-1} G\left(\frac{p}{2N}\right) e^{-2\pi i pk/2N} + G\left(\frac{1}{2}\right) \right]$$

The corresponding formula for the midpoint integration method is

$$A_k = \frac{1}{N} \sum_{p=0}^{N-1} G\left(\frac{p + 1/2}{2N}\right) e^{-(2\pi i/2N)(p+1/2)k}$$

We shall see these formulas in a different situation in the next chapter.

Exercise

10.5-1   Work out the formulas of Section 10.5 for the odd transfer function $\tilde{H}(f) = -\tilde{H}(-f)$.

10.6   ANOTHER METHOD OF DESIGN

The coefficients a_k and b_k of the transfer function are sometimes hard to obtain, and the finite Fourier series provides an alternate approach. In the band $0 \leq f \leq \frac{1}{2}$ we select the $N + 1$ points

$$ x_p = \frac{p}{2N} \qquad (p = 0, 1, \ldots, N) $$

and find either the sine or the cosine expansion of the given transfer function, depending on its oddness or evenness. The fast Fourier transform gives the finite Fourier series expansion coefficients (the capital letter coefficients), which include the aliasing that arises from the finiteness of the discrete expansion of the transfer function (this aliasing is not to be confused with that from the original sampling of the signal).

10.7   COMPUTATION OF POWER SPECTRA

The main use of the fast Fourier transform is to compute power spectra. For a real function, the complex Fourier expansion has coefficients with the property

$$ c_k = \overline{c_{-k}} $$

and so the power spectrum has the values

$$ |c_k|^2 = c_k c_{-k} $$

Just what comes out of the local library package requires careful reading of the description.

Frequently the local library package requires an exact power of 2 for the number of data points, and usually you do not have that many. There is a great temptation simply to pad out the data with a string of zeros. This step should be resisted. And even if you do have exactly the right number, remember that the Fourier expansion implies that the function is periodic; if there is a significant difference between the starting values and the ending values, there will be a discontinuity in the function and you will see the effect that discontinuity in the spectrum.

Conventional wisdom these days suggests first removing the mean of the data (but see Section 13.4) and then using the set of weights

$$w(k) = \frac{1 + \cos{(\pi k/N_1)}}{2}$$

on the data, where N_1 is about 10% of the data that you have. This operation produces a taper at the starting end. A similar (but reversed) sequence of weights at the other end produces a second taper. Then you can pad out the rest with zeros. Thus the flat part of the window is not more than about 80% of the data, and there are no sharp corners in the data going into the fast Fourier transform. Whether you pad at both ends or only one end of the data is a matter of phase shift only. See Fig. 10.7-1.

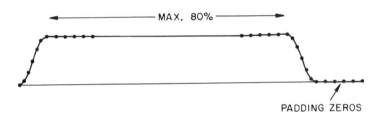

FIGURE 10.7-1   ADJUSTING DATA TO FIT THE FAST FOURIER TRANSFORM

If you have exactly the right amount of data and merely face a discontinuity around the ends, a taper of about 10% of the data to smooth the transition is suitable.

11

Recursive Filters

11.1 WHY RECURSIVE FILTERS?

In Section 1.1 we defined a recursive filter by the formula

$$y_n = \sum_{k=-K}^{K} c_k x_{n-k} + \sum_{k=1}^{K} d_k y_{n-k}$$

where in the second sum the $k = 0$ term is omitted. It is assumed, of course, that not all the d_k are zero; otherwise we would have a simple nonrecursive filter.

A recursive filter has more "memory" than a nonrecursive filter (all the $d_k = 0$); a nonrecursive filter can respond only to the values of x_m in the range x_{n-K} to x_{n+K}. An example of a recursive filter is a formula for integration that *must* "remember" all the past values back to and including the value at the lower limit of the integration. In Section 3.4 we examined the trapezoid rule for integration

$$y_{n+1} = y_n + \frac{h}{2}[x_{n+1} + x_n] \qquad y_0 = \text{constant}$$

which is clearly a recursive filter. Similarly, Simpson's formula

$$y_{n+1} = y_{n-1} + \frac{h}{3}[x_{n+1} + 4x_n + x_{n-1}] \qquad y_0 = \text{constant}$$

is a recursive filter.

Even a short "span" of coefficients in a recursive filter can remember all the past. In the simplest case,

$$y_{n+1} = ay_n + x_n \qquad y_0 = 0$$

where all the x_n are zero except $x_0 = 1$ and all the y_k for negative indices are zero, we get

$$y_1 = 1, \quad y_2 = a, \quad y_3 = a^2, \quad y_4 = a^3, \ldots, y_n = a^{(n-1)}$$

Thus we have a geometric progression of values for y_n, and the value of x_0 persists indefinitely. This ability to remember values that occurred long ago represents a definite asset of the recursive filter in many situations.

Many recursive filters are used in "real-time" problems, meaning that values of the data x_m and output y_m for indices $m > n$ are *not* available for use at the time that y_n is being computed. Such filters are often called (Section 1.1) *physically realizable* as contrasted with the typical nonrecursive filter, which is *physically unrealizable*, meaning that when we calculate a value y_n, we are using data x_m for $m > n$ from the future. Although physically non-realizable recursive digital filters are possible, they occur so seldom that we will concentrate on recursive filters of the form

$$y_n = \sum_{k=0}^{M} c_k x_{n-k} + \sum_{k=1}^{N} d_k y_{n-k}$$

(but see Section 11.6). Although "one-sided" physically realizable filters, we will use the name recursive filters for them. Sometimes they are called "infinite impulse response" filters from their ability to produce, from a single impulse, effects arbitrarily far into the future. They are often abreviated as IIR filters. Correspondingly, nonrecursive filters are abreviated FIR, "finite impulse response." Degenerate examples of recursive filters that do not continue an impulse indefinitely may be constructed but will not be considered here.

Recalling that for a linear system the output frequency must be the same as the input frequency (Section 2.5), we make our usual substitutions

$$x_n = A_1 e^{2\pi i f n} \qquad y_n = A_0 e^{2\pi i f n}$$

and obtain

$$\tilde{H}(f) = \frac{A_0}{A_1} = \frac{\sum_{k=0}^{M} c_k e^{-2\pi i f k}}{1 - \sum_{k=1}^{N} d_k e^{-2\pi i f k}}$$

as our transfer function $\tilde{H}(f)$. Thus the transfer function of a recursive filter is a rational function in $e^{-2\pi i f}$.

We are now in a position to see why a recursive filter can have a second important property—namely, that the transition band from pass to stop can be narrow. This situation follows from the observation that when the polynomial in the denominator goes near zero, the quotient can change rapidly, can rise sharply, and allow a narrow transition band in a recursive filter.

11.2   REDUCTION TO SIMPLER FORM

Recall that in order to get smooth filters (Chapter 7), we made a transformation that converted the problem into one that was equivalent to fitting a polynomial to the transfer function characteristic that we wanted (Section 7.2). In the recursive filter case, we do not have a polynomial but rather a rational function in both sines and cosines. From calculus, we remember that a rational function in both sines and cosines can be reduced to a rational function in w by the "tangent half-angle substitution"

$$\tan \frac{\omega}{2} = \tan \frac{2\pi f}{2} = w$$

Before using this transformation blindly, let us pause and look a little closer at how and why it works. Actually, we are in the position of having a rational function in $e^{2\pi i f}$.

When we have a polynomial in w, the most general "one-to-one" substitution that we can make for w and still preserve the polynomial form of the same degree is a linear one

$$w' = aw + b, \qquad w = \frac{1}{a}(w' - b)$$

However, with a rational function we have an increased range of substitutions that will not alter the basic form and complexity; we can now use the linear fractional transformation

$$w' = \frac{aw + b}{cw + d} \qquad w = \frac{dw' - b}{a - cw'}$$

Notice that one of the coefficients can be taken to be 1.

We first ask the question: Which transformations in this three-parameter family

$$e^{2\pi i f} = \frac{aw + b}{cw + d}$$

will have the property that for real f we get real w? We can determine the details of the transformation as follows.

To fix the "phase" of f (which otherwise would be arbitrary and only add to the confusion), we can require

$$f = 0 \text{ corresponds to } w = 0$$

This means

$$e^0 = 1 = \frac{b}{d} \quad \text{or} \quad b = d$$

Since one coefficient is arbitrary, we can set

$$b = d = 1$$

We now have

$$e^{2\pi i f} = \frac{1 + aw}{1 + cw}$$

To determine the "direction" of the transformation, we can require

$$f \longrightarrow \frac{1}{2} \text{ corresponds to } w \longrightarrow +\infty$$

This means that at $f = \frac{1}{2}$ we have

$$e^{\pi i} = -1 = \frac{a}{c} \quad \text{or} \quad c = -a$$

We now have

$$e^{2\pi i f} = \frac{1 + aw}{1 - aw}$$

Finally, to fix the "scale" of the transformation, we can require

$$f = \frac{1}{4} \text{ corresponds to } w = 1$$

This means

$$e^{\pi i/2} = i = \frac{1 + a}{1 - a}$$

Multiply both sides by $(1 - a)$ to obtain

$$i - ia = 1 + a$$

which implies

$$a = i$$

We have finally determined the transformation completely (using the three conditions)

$$e^{2\pi i f} = \frac{1 + iw}{1 - iw}$$

Notice that for real f and real w each side has absolute value 1.

Writing this formula in the "real" form

$$\cos 2\pi f + i \sin 2\pi f = \frac{1 + 2iw - w^2}{1 + w^2}$$

and equating real and imaginary parts gives the usual "tangent half-angle" formulas of the calculus

$$\cos 2\pi f = \frac{1 - w^2}{1 + w^2} \qquad \sin 2\pi f = \frac{2w}{1 + w^2}$$

As stated in the first paragraph of this section,

$$\tan \pi f = \tan\left[\frac{2\pi f}{2}\right] = \frac{1 - \cos 2\pi f}{\sin 2\pi f} = \frac{1 + w^2 - (1 - w^2)}{2w} = w$$

Using this transformation, our transfer function becomes a rational function in w with complex coefficients.

Before continuing, let us examine the transformation, Fig. 11.2-1. As the original variable f went from 0 to $\frac{1}{2}$, the new variable w goes from 0 to infinity. Because of symmetry, as f goes from 0 to $-\frac{1}{2}$, w goes from 0 to minus infinity.

For our transfer function of the recursive filter, we now have a rational function in w, with w real, but we still have complex coefficients. It is con-

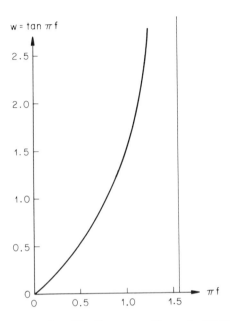

FIGURE 11.2-1   THE FREQUENCY TRANSFORMATION

ventional to ignore the effects due to the imaginary terms, the phase relation (for each frequency) between the input and the output of the filter; in
mathematical words, we ignore the actual complex value of the ratio of
A_0/A_I and concentrate on its absolute value. This step is easiest done by
taking the product of $\tilde{H}(f)$ times its conjugate. The product is, of course,
real. Since i occurs only with f in the expressions, the procedure is the same
as using

$$\tilde{H}(f)\tilde{H}(-f)$$

To summarize, we have replaced the original problem—that of finding
a recursive filter to fit a given transfer function in the Nyquist interval—with
the problem of matching the square of the modulus of the transfer function
along the whole real axis w. The latter *omits all information about the phase
relationships* between the input and output signal frequencies (but see Section
11.6). Using the tangent half-angle transformation, we now have a problem
of fitting a rational function in w, over the range minus infinity to infinity,
to the square of the modulus of the transfer function (transformed, of course,
to the w variables).

11.3   STABILITY AND THE Z TRANSFORMATION

Before going deeper into the design problem, let us look at the problem of undoing the replacement of the original problem of matching $\tilde{H}(f)$ by the problem of matching the product

$$\tilde{H}(f)\tilde{H}(-f)$$

This replacement has doubled the degree of the rational function and, in particular, has doubled the number of zeros in the denominator. As is always the case in computing when using old values, we need to consider the *stability* of the digital filter. By "stability" we usually mean that a bounded sequence of input values will produce a bounded output, although not necessarily the same bound. Actually, we will allow a polynomial growth to the output. In short, we must return to our original difference equation and see what restrictions must be placed on it in order to achieve this stability. When we know, we will have some idea of how to go from the new problem back to our original problem.

We write our difference equation in the form

$$y_n = \sum_{k=1}^{N} d_k y_{n-k} + \sum_{k=0}^{M} c_k x_{n-k}$$

The x_{n-k} are the given data, and so the last sum can be regarded as a known function of n. We are therefore reduced to studying the simple linear difference equation in y_n with a forcing function (the second sum). The coefficients of the difference equation are constants, and its theory corresponds closely to that of linear differential equations with constant coefficients. In the theory of differential equations with constant coefficients, we guessed a solution of the form

$$e^{mt}$$

and by substituting this solution into the differential equation, we were immediately led to the *characteristic equation*

$$m^N = \sum_{k=1}^{N} d_k m^{N-k}$$

For the corresponding situation in linear difference equations, we guess at a solution of our difference equation of the form

$$\rho^m$$

By direct substitution, we are led to the corresponding characteristic equation

$$\rho^N = \sum_{k=1}^{N} d_k \rho^{N-k}$$

In general, this Nth-degree polynomial in ρ will have N distinct zeros, and to each there will be a term in the general solution of the homogeneous equation, of the form

$$C_k \rho_k^m$$

To a linear combination of these N solutions of the homogeneous equation, we add any one particular solution of the complete equation (include the forcing term), and we have the complete solution of the original nonhomogeneous linear difference equation. For a multiple root of multiplicity j, we try the corresponding functions ρ^m, $m\rho^m$, $m^2\rho^m$, ..., $m^{j-1}\rho^m$ to get the corresponding j linearly independent solutions.

The initial conditions determine the coefficients C_k of the individual solutions of the homogeneous equation, and, in principle, they could make one or more of the solutions disappear. In actual computation, because of round-off if no other reason, all the solutions will sooner or later appear. If we want a stable solution (see above), then we must require that *none* of the zeros of the characteristic equation has a magnitude greater than 1 in size. If there were one or more such solutions, clearly the total solution would eventually grow exponentially large in size and the filter would be unstable. It is unlikely, although not impossible, that such a filter would be useful over a long stretch of data (it could be useful over a small piece), and so we will impose the restraint on our design that the filter be stable—that is, that the roots of the characteristic equation have a size less than 1 or at most 1 in certain selected cases. The admission of 1 as a root puts the filter on the margin of stability, as in both the trapezoid and Simpson's integration formulas. If a multiple root of size 1 exists, we will have a polynomial type of growth in the stable solution.

We used p a moment ago because it is conventional notation in the field of difference equations, but in recursive filter theory it is customary to use z, where

$$z = e^{2\pi i f}$$

Since $|z|$ is no longer necessarily on the unit circle, the variable f is not necessarily real.

We are looking for solutions of the difference equation of the form

$$z^m$$

The characteristic polynomial of the difference equation of the filter is then the polynomial

$$z^N - \sum_{k=1}^{N} d_k z^{N-k} = 0$$

and the condition that no solution of this polynomial be greater than 1 in size means that no solution of the homogeneous equation can grow exponentially. The classical filter theory at this point requires that all the zeros be less than 1; but since we are interested in numerical integration, among other things, we permit a zero on the boundary of the circle. For second-order linear differential equations without a first derivative present, we will have to allow a double zero (which leads to a solution of the homogeneous difference equation that grows like n, the number of steps integrated).

The solution of the homogeneous equation is often called "the transient" solution. Stated simply, the conventional theory requires that the transient die out sufficiently far away, but we are permitting it to persist (so we can remember the initial conditions)—and, indeed, in some cases allow polynomial growth—but we still avoid exponential growth.

The transformation to the w variable is

$$z = \frac{1 + iw}{1 - iw} \quad \text{or} \quad w = i\left[\frac{1 - z}{1 + z}\right]$$

In the w notation, the stability condition is

$$|z| = \left|\frac{1 + iw}{1 - iw}\right| \leq 1$$

In applying this condition, we recall that z goes around the unit circle in the complex plane from $-\pi$ to $+\pi$. On the circle we have $z = e^{2\pi i f}$ for real f. Correspondingly w goes along the real axis from $-\infty$ to $+\infty$, and we see that the inside of the unit circle in the z plane is mapped into a half-plane in w. To decide which half, we simply find where $z = 0$ goes. This point is $w = i$. Thus the inside of the circle $|z| = 1$ goes into the upper half of the w plane. For our filters we will therefore require that the poles of the transfer function (the zeros of the denominator) all lie in, or on the edge of, the upper half plane in w. (Some textbooks use a mapping of z to w such that it is the right or left half-plane that can be used—which is only a question of an extra multiplier of i in the transformation.)

The problem of selecting which zeros of the denominator to use for a stable filter arises from the fact that we were given a transfer function $\tilde{H}(f)$, and to ensure that it was real, we started our design by approximating

$$\tilde{H}(f)\tilde{H}(-f)$$

Since f only occurs in conjunction with i, this expression is equivalent, as we saw, to the product of the transfer function times its complex conjugate. When we finally derive the approximation to this real function, we face the messy problem of extracting a suitable $\tilde{H}(f)$ that will be stable (zeros all in the upper half-plane).

11.4 BUTTERWORTH FILTERS

Perhaps the most famous recursive filter design is the Butterworth filter. Such filters correspond to our earlier smooth nonrecursive filters (Chapter 7). For the Butterworth filter we pick the rational function

$$\tilde{H}(f)\tilde{H}(-f) = \frac{1}{1 + [w/w_c]^{2N}}$$

This function has the simple features of high tangency at both the origin and infinity (in the w variable) and smoothness elsewhere.

The design features are shown in Fig. 11.4-1. Based on this figure, we can obtain the design parameters, N and w_c, as follows. First, we write the two conditions that arise at the edges of the transition band

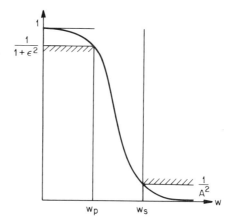

FIGURE 11.4-1   BUTTERWORTH FILTER

$$\frac{1}{1+\epsilon^2} = \frac{1}{1+[w_p/w_c]^{2N}}$$

$$\frac{1}{A^2} = \frac{1}{1+[w_s/w_c]^{2N}}$$

Next, take the reciprocal of both equations, do some simplification, and get

$$\left[\frac{w_p}{w_c}\right]^{2N} = \epsilon^2$$

$$\left[\frac{w_s}{w_c}\right]^{2N} = A^2 - 1$$

To eliminate w_c, divide one equation by the other

$$\left[\frac{w_p}{w_s}\right]^{2N} = \frac{\epsilon^2}{A^2 - 1}$$

Solving for N, we have

$$N = \frac{\log\left(\epsilon/\sqrt{A^2-1}\right)}{\log\left(w_p/w_s\right)}$$

and hence

$$w_c = \frac{w_p}{(\epsilon)^{1/N}}$$

Thus we determine the product $\tilde{H}(f)\tilde{H}(-f)$.

To find a suitable $\tilde{H}(f)$, we proceed as follows. The poles of the expression are given by the zeros of the denominator—that is, by

$$1 + \left[\frac{w}{w_c}\right]^{2N} = 0$$

$$\left[\frac{w}{w_c}\right]^{2N} = -1 = e^{\pi i + 2\pi i k}$$

$$\frac{w}{w_c} = e^{\pi i (2k+1)/2N} \qquad (k = 0, 1, \ldots, 2N - 1)$$

The zeros that we want are in the upper half-plane and are clearly given by

$$k = 0, \ldots, N - 1$$

If we pair these zeros off one from each end—pair k with $N - k - 1$—we obtain for each pair

$$\left[\frac{w}{w_c} - e^{\pi i (2k+1)/2N}\right]\left[\frac{w}{w_c} - e^{\pi i (2N - 2k - 1)/2N}\right]$$

Since

$$e^{\pi i (2N/2N)} = e^{\pi i} = -1$$

this becomes

$$\left[\frac{w}{w_c} - e^{\pi i (2k+1)/2N}\right]\left[\frac{w}{w_c} + e^{-\pi i (2k+1)/2N}\right]$$

$$= \left[\frac{w}{w_c}\right]^2 - 2i \sin\left\{\frac{\pi(2k + 1)}{2N}\right\}\left[\frac{w}{w_c}\right] - 1$$

To get back to the z variable, we substitute

$$w = i\left[\frac{1 - z}{1 + z}\right]$$

and we have

$$\left[\frac{-(1/w_c^2)(1 - z)^2 + 2 \sin\{\pi(2k + 1)/2N\}[1 - z^2][1/w_c] - (1 + z)^2}{(1 + z)^2}\right]$$

as the real quadratic factor corresponding to the two complex, conjugate linear factors. There are, of course, at most $N/2$ of these factors, one for each k, $0 \leq k \leq (N/2)$.

If N is odd, then there is one factor left

$$\left[\frac{w}{w_c} - e^{\pi i/2}\right] = \left[\frac{w}{w_c} - i\right]$$

and the substitution back to z gives

$$i\left[\frac{(1/w_c)(1-z) - (1+z)}{1+z}\right]$$

These factors were in the denominator; thus inverting them gives

$$\tilde{H}(f) = \prod \frac{(1+z)^2}{\{\text{quadratic factors}\}}$$

with a possible linear factor included. The construction of a polynomial from its zeros leaves the multiplier of the whole polynomial arbitrary. We use the fact that at $f = 0$ we want unity gain in the filter, which means that we must drop the multiplier of -1 in each quadratic factor and i in the linear factor *when* it occurs in the reconstruction of the polynomial from its zeros.

It might be supposed that at this point we would multiply out the numerator and denominator to find the coefficients of the individual powers of z and thus obtain the coefficients of the filter. However, doing so is unnecessary *and* can lead to serious roundoff errors. Instead we will simply pair off a factor of degree 2 in the numerator with each quadratic factor in the denominator and one of degree 1 if there is a final linear factor in the denominator. In this form we have the desired filter transfer function characteristic $\tilde{H}(f)$ as a cascade (product) of second-order transfer functions (possibly with one first-order)

$$\tilde{H}_c(f) = \frac{w_c^2[z^2 + 2z + 1]}{z^2[w_c^2 - 2w_c \sin\{\pi(2k+1)/2N\} - 1]} \\ + z[2 + 2w_c^2] + [w_c^2 + 2w_c \sin\{\pi(2k+1)/2N\} - 1]$$

where

$$z = e^{2\pi i f}$$

To get back to the original digital filter, recall that the second-order recursive digital filter

$$y_n = c_0 x_n + c_1 x_{n-1} + c_2 x_{n-2} + b_1 y_{n-1} + b_2 y_{n-2}$$

has (Section 11.1) the transfer function

$$\tilde{H}_c(f) = \frac{c_0 + c_1 z^{-1} + c_2 z^{-2}}{1 - b_1 z^{-1} - b_2 z^{-2}} = \frac{c_0 z^2 + c_1 z + c_2}{z^2 - b_1 z - b_2}$$

Thus we can identify the coefficients of the transfer function with the coefficients of the digital filter that we want: the process requires noting that the coefficient of z^2 in the denomination of the second form is 1, and so in the first form we must divide each coefficient by the coefficient of z^2 before equating the corresponding coefficients of the two forms.

Our original data x_n is now processed by one second-order filter; the output is processed by the next filter, and so on, perhaps including one first-order filter. In this way, we have the results wanted, the original x_n processed by the nth-order Butterworth filter.

11.5 A SIMPLE CASE OF THE BUTTERWORTH
   FILTER DESIGN

To illustrate the design of a Butterworth filter, we select the simplest example (smallest N) of a Butterworth filter that illustrates the method adequately and carry out the numerical details carefully, following the steps in the previous section. Let $N = 3$. First, we find the zeros of

$$1 + \left(\frac{w}{w_c}\right)^6 = 0$$

which are

$$\frac{w}{w_c} = e^{\pi i(1 + 2k)/6}$$

Only for $k = 0, 1, 2$ does the angle satisfy

$$0 \le \frac{\pi(1 + 2k)}{6} \le \pi$$

which is the condition for the value to be in the upper half-plane. We pair the cases $k = 0$ and 2 to form the quadratic factor

$$\left[\frac{w}{w_c} - e^{\pi i/6}\right]\left[\frac{w}{w_c} - e^{5\pi i/6}\right] = \left[\frac{w}{w_c} - e^{\pi i/6}\right]\left[\frac{w}{w_c} + e^{-\pi i/6}\right]$$

$$= \left(\frac{w}{w_c}\right)^2 - \frac{w}{w_c}[e^{\pi i/6} - e^{-\pi i/6}] - 1$$

$$= \left(\frac{w}{w_c}\right)^2 - 2i\frac{w}{w_c}\sin\frac{\pi}{6} - 1$$

But $\sin \pi/6 = 1/2$, and so we have

$$\left(\frac{w}{w_c}\right)^2 - i\frac{w}{w_c} - 1$$

The remaining, unpaired factor comes from $k = 1$

$$\frac{w}{w_c} - e^{\pi i/2} = \frac{w}{w_c} - i$$

We convert these factors from w to z, using

$$w = \frac{i(1-z)}{1+z}$$

and get (dropping the multiplicative factor of -1 in the quadratic term and i in linear term so that at $w = 0$ we have unity gain)

$$\frac{1}{\left[\left(\frac{1-z}{1+z}\right)^2\frac{1}{w_c^2} - e^{\pi i}\left(\frac{1-z}{1+z}\right)\frac{1}{w_c} + 1\right]\left[\left(\frac{1-z}{1+z}\right)\frac{1}{w_c} - 1\right]}$$

$$= \left[\frac{(1+z)^2 w_c^2}{z^2(1+w_c^2 - w_c) + 2z(w_c^2 - 1) + (1 + w_c^2 + w_c)}\right]\left[\frac{(1+z)w_c}{1 - w_c - z(1 + w_c)}\right]$$

$$= \tilde{H}(f)$$

which is the product of a second-order section and a first-order section filter. These two simple filters are then used to process the original signal x_n.

11.6   REMOVING THE PHASE: TWO-WAY FILTERS

Butterworth filters (and most other recursive filters) have, since they are not symmetric, a phase relation between the input and output signals that is not the same for all frequencies. Often this phase is annoying, but *if* we can process the data in both a forward and backward direction, then there is a simple solution to the problem of eliminating the phase. We merely process the data by the linear filter and run this output taken in the reverse direction through the same filter. If there is a phase shift at a given frequency in the first pass through the filter, there is the same phase shift of the opposite sign at this same frequency in the second pass. Because we are processing the output of the first pass in the reverse direction, the two phase shifts must exactly cancel. We, of course, obtain the *square* of the absolute value of the transfer function as the effective transfer function. Thus if we wish to use this trick, then in our original design we allow for this squaring of the transfer function that arises from the two passes.

The trick is so simple that it is apt to be overlooked. The reader is reminded especially to keep it in mind for those situations in which the data may be processed in either of the two directions. Typically, this procedure can be done if the data is fully recorded before the processing starts. It clearly cannot be done in real-time data processing. Remember that the double pass will square the transfer function.

12

Chebyshev Approximation and Chebyshev Filters

12.1 INTRODUCTION

The problem of measuring how close a designed filter is to the original ideal filter that we started with occurs constantly. We have used the measure of least squares, modifying it whenever we applied a window; in this chapter we introduce another criterion of closeness, *the Chebyshev criterion*, which uses the maximum difference between two curves as the measure of the distance between them.

The Chebyshev criterion is very popular these days, not only in filter theory where we wish to know the worst that we have done but also in many situations in computing where, for instance, we wish to know that a certain library routine gives answers "at least as good as . . . ." In finding a Chebyshev fit, we adjust the parameters to make the maximum error as small as possible: we *minimize the maximum error*, and so the process is often called *the minimax strategy* of fitting.

The price to be paid for this fit, as compared to least squares, is that the sum of the squares of the errors is larger for a Chebyshev fit than for the corresponding least-squares fit; but also, conversely, the least-squares fit has a larger maximum error than does the Chebyshev fit. In filter design the Chebyshev fit is frequently preferred. Almost always the Chebyshev criterion distorts the least-squares fit less than least squares distorts the Chebyshev.

In the Chebyshev approximation we usually have an error curve that turns out to have equal-sized ripples. The reason is intuitively evident. Imagine that you have some kind of fit to the data and that you are looking at the error curve. If you try to push down the worst peak of the error curve by a change of the filter coefficients, then some other place will naturally increase in size a little. If at any stage you try to push down all the worst extremes at the same time, you will (probably) be able to do so until there are more maximum-sized extremes than you have adjustable parameters, whereupon you can go no farther. Thus the Chebyshev approximation is often called *the equal ripple approximation.*

Our approach to the design of Chebyshev filters will be via the Chebyshev polynomials, which individually have the equal ripple property and are therefore closely related to the equal ripple fitting process. Finally, we will look at the Chebyshev design of a simple integrator.

12.2 CHEBYSHEV POLYNOMIALS

The Chebyshev polynomials are defined by

$$T_n(x) = \cos(n \arccos x) \quad (n = 1, 2, \ldots)$$
$$T_0(x) = 1$$

As the definition stands, $T_n(x)$ does not look like a polynomial of degree n. However, in Section 9.2 we showed that $\cos nx$ is a polynomial in $\cos x$ of exactly degree n. We now see, since $\cos(\arccos x) = x$, that our definition of the Chebyshev polynomial $T_n(x)$ does indeed define a polynomial of degree n.

The Chebyshev polynomials are closely related to the cosine functions, as can be seen from the following change of variable

$$x = \cos \tau$$

or, equivalently,

$$\tau = \arccos x$$

See Fig. 12.2-1 for what the tranformation does in detail, how it stretches the independent variable when going from one notation to the other (equally spaced τ do not give equally spaced x). It is convenient to define the polynomials for negative indices by the obvious formula

$$T_{-k}(x) = T_k(x)$$

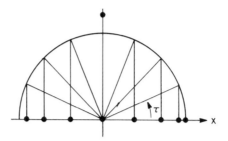

FIGURE 12.2-1

From the fact that the cosines are orthogonal functions (over both the continuous interval and discrete sets of points), we see that the continuous Chebyshev polynomials are also orthogonal with some weight function over the interval $-1 \le x \le 1$. To find the details, we start with the orthogonality conditions of Section 4.4 (note that as a result of Section 10.2 there are also discrete versions corresponding to the continuous one that we are using)

$$\int_0^\pi \cos m\tau \cos n\tau \, d\tau = \begin{cases} 0, & m \ne n \\ \dfrac{\pi}{2}, & m = n \ne 0 \\ \pi, & m = n = 0 \end{cases}$$

Changing the variable from τ to x, using $\tau = \text{arc cos } x$, we get

$$\int_{-1}^1 \frac{T_m(x)T_n(x)}{\sqrt{1-x^2}} \, dx = \begin{cases} 0, & m \ne n \\ \dfrac{\pi}{2}, & m = n \ne 0 \\ \pi, & m = n = 0 \end{cases}$$

From this result we see that the Chebyshev polynomials are orthogonal over the interval $-1 \le x \le 1$ with weight function

$$\frac{1}{\sqrt{1-x^2}}$$

For the discrete set of points, the unequal spacing of the sample points plays the role of the weight function in the summation over the sample points to produce the orthogonality.

From the trigonometric identity

$$\cos (n + 1)\tau + \cos (n - 1)\tau = 2 \cos \tau \cos n\tau$$

we obtain, by changing variables from τ to x, the three-term recurrence relation

$$T_{n+1}(x) - 2xT_n(x) + T_{n-1}(x) = 0$$

It is now easy to generate the first few Chebyshev polynomials [clearly, $T_0(x) = 1$ and $T_1(x) = x$]

$$T_2(x) = 2x^2 - 1$$
$$T_3(x) = 4x^3 - 3x$$
$$T_4(x) = 8x^4 - 8x^2 + 1$$
$$\cdots$$

An examination of these polynomials and the three-term relation shows that the nth polynomial has a leading coefficient of

$$2^{n-1}$$

and that the polynomials are alternately odd and even functions of x. Examination also shows that since $T_0(1) = T_1(1) = 1$, it follows that $T_n(1) = 1$ for all n.

Exercises

12.2-1   Prove $T_n(-1) = (-1)^n$.

12.2-2   Compute $T_5(x)$ and $T_6(x)$.

12.2-3   Show that $T_{2n}(x) = T_n(2x^2 - 1) = 2T_n^2(x) - 1$.

12.2-4   Show that $T_n[T_m(x)] = T_{mn}(x)$.

12.3   THE CHEBYSHEV CRITERION

The theorem that we wish to prove in this section, and the one that connects Chebyshev polynomials with the minimax criterion, is that of all polynomials of degree n with leading coefficient 1 the Chebyshev polynomial divided by 2^{n-1} has the least maximum in the interval $-1 \leq x \leq 1$ [$T_n(x)/2^{n-1}$ is the "smallest" polynomial of degree n beginning $x^n + \cdots$].

The proof is fairly easy. To start, we note that the nth Chebyshev polynomial, being a cosine in disguise, has the equal ripple property and has exactly $n + 1$ extreme values in the interval. If there were another polynomial, say $\phi_{n-1}(x)$, with smaller extreme values, then the difference

$$\frac{T_n(x)}{2^{n-1}} - \phi_{n-1}(x) = \text{a polynomial in } x \text{ of degree at most } (n - 1)$$

We now consider the value of this expression at the $(n + 1)$ extreme values of the nth Chebyshev polynomial. Clearly, by the definition of $\phi_{n-1}(x)$, the values of this difference will be alternately positive and negative at successive extremal points. (See Fig. 12.3-1.) Thus there will be at least n changes of sign for the difference function that is a polynomial of degree at most $(n - 1)$, a contradiction! We conclude, therefore, that the Chebyshev polynomial divided by 2^{n-1} has the smallest value in the interval of any polynomial of that degree with leading coefficient 1.

This discussion shows the central role of the Chebyshev polynomials when we are interested in the minimax (Chebyshev) criterion for selecting a polynomial fit. If the error curve can be written as a simple Chebyshev polynomial, then we will have the best Chebyshev fit. If we can represent a function as a series in Chebyshev polynomials and truncate the series, the error will look like the first neglected term, *provided* that the series is reasonably rapidly convergent.

The fact that the Chebyshev polynomials are also orthogonal polynomials means, as we have seen, that they give the best least-squares fit (relative to their weight function) in the interval. That they also give the minimax fit

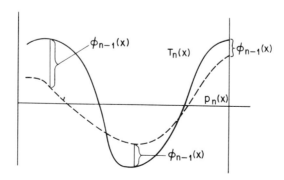

FIGURE 12.3-1

seems to be a contradiction. It is the weight function

$$\frac{1}{\sqrt{1 - x^2}} \qquad (-1 \le x \le 1)$$

which puts great weight at the ends of the interval where the least-squares fit tends to be worst, that explains how an expansion in Chebyshev polynomials can give both the best-weighted least-squares fit and *almost* the best minimax fit (the error is actually not the first term neglected but the sum of all the neglected terms). Note that an expansion in Chebyshev polynomials is a Fourier cosine expansion in disguise.

12.4 CHEBYSHEV FILTERS

In the chapters on nonrecursive filters we saw the advantage, in terms of a narrower transition band, and of allowing ripples in either the pass or stop bands of a filter, or in both. So we now turn to lowpass *recursive* filters in which we allow ripples in either the pass or stop band.

The criterion used in our design is the Chebyshev one of the minimum-maximum error—the equal ripple approach. This choice is quite natural in view of the use that we usually make of a digital filter; we often want to control the *maximum* error.

The design method is, as in the case of the Butterworth filters, "Behold! this formula does the trick." Thus we need only observe that the proposed formula has the properties that it is supposed to have; no elaborate theory is necessary to deduce the design or to use later if something different but quite similiar to the standard design objective is wanted.

12.5 CHEBYSHEV FILTERS, TYPE 1

The first design, type 1, allows ripples in the pass band (Fig. 12.5-1). To meet this design, we start with (in the polynomial variable w)

$$\tilde{H}(f)\tilde{H}(-f) = \frac{1}{1 + \epsilon^2 [T_n(w/w_p)]^2}$$

where $T_n(x)$ is the Chebyshev polynomial of order n. We see immediately

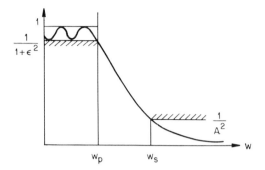

FIGURE 12.5-1 CHEBYSHEV TYPE 1 FILTER

that as long as $|w| \leq |w_p|$, we have

$$\frac{1}{1+\epsilon^2} \leq |\tilde{H}(f)\tilde{H}(-f)| \leq 1$$

and at $w = w_p$ we have the equal ripple Chebyshev polynomial beginning to break out of the limits $-1 \leq T(w/w_p) \leq 1$.

The fact that the Chebyshev polynomial in the interval $-1 \leq x \leq 1$ is the smallest of any polynomial of the same degree and same leading coefficient means, conversely, that given the bound, it rises most rapidly outside (we are sure of this fact if we go far enough away from 1, but it is farily obvious that it applies near 1).

Recall that

$$\sin x = \frac{e^{ix} - e^{-ix}}{2i} \qquad \sinh x = \frac{e^x - e^{-x}}{2}$$

$$\cos x = \frac{e^{ix} + e^{-ix}}{2} \qquad \cosh x = \frac{e^x + e^{-x}}{2}$$

By direct substitution, we see immediately that

$$\sin ix = i \sinh x \qquad \sinh ix = i \sin x$$
$$\cos ix = \cosh x \qquad \cosh ix = \cos x$$

Therefore the analytic representation of the Chebyshev polynomials over all x values is

$$T_n(x) = \begin{cases} \cos\,[n \arccos x], & |x| \leq 1 \\ \cosh\,[n \operatorname{arccosh} x], & |x| \geq 1 \end{cases}$$

and when $|x| > 1$, we expect a rapid growth in $T_n(x)$ that will make the transition band from "pass" to "stop" fairly narrow.

Returning to the transfer function, we see that at the edge of the pass band, $w = w_p$, we have $T_n(1) = 1$, and the condition at the edge of the pass band is satisfied.

The condition at the edge of the stop band, where $w = w_s$, is

$$\frac{1}{1 + \epsilon^2 T_n^2(w_s/w_p)} = \frac{1}{A^2}$$

or

$$\left| T_n\left(\frac{w_s}{w_p}\right) \right| = \frac{\sqrt{A^2 - 1}}{\epsilon}$$

Hence

$$n = \frac{\text{arccosh} \left[\sqrt{A^2 - 1}/\epsilon \right]}{\text{arccosh} \left(w_s/w_p \right)}$$

and we have the "design parameter" n.

There is an alternate representation of the arccosh x. If

$$y = \text{arccosh } x$$

then

$$x = \cosh y = \frac{e^y + e^{-y}}{2}$$

This result can be written as

$$(e^y)^2 - 2x(e^y) + 1 = 0$$

or

$$e^y = x + \sqrt{x^2 - 1}$$

where we have used the $+$ sign for $|x| > 1$. Thus

$$y = \text{arccosh } x = \ln \left[x + \sqrt{x^2 - 1} \right]$$

As with the Butterworth filters, in order to obtain the transfer function $\tilde{H}(f)$ from $\tilde{H}(f)\tilde{H}(-f)$, we must find the zeros of the denominator

$$1 + \epsilon^2 T_n^2\left(\frac{w}{w_p}\right) = 0$$

or

$$T_n\left(\frac{w}{w_p}\right) = \pm\frac{i}{\epsilon}$$

Let $(w/w_p) = x$

$$\cos\left[n \arccos x\right] = \pm \frac{i}{\epsilon}$$

$$n \arccos x = \arccos\left(\pm\frac{i}{\epsilon}\right)$$

To find the value of an arccos of an imaginary number, set

$$\arccos\left(\pm\frac{i}{\epsilon}\right) = \alpha + i\beta$$

$$\pm\frac{i}{\epsilon} = \cos(\alpha + i\beta)$$

$$\pm\frac{i}{\epsilon} = \cos\alpha\cosh\beta - i\sin\alpha\sinh\beta$$

Equating real and imaginary parts gives the two equations

$$\cos\alpha\cosh\beta = 0$$

$$\sin\alpha\sinh\beta = \mp\frac{1}{\epsilon}$$

From the first of these two equations, since $\cosh\beta \neq 0$ (real β),

$$\cos\alpha = 0$$

and therefore

$$\alpha = \frac{\pi}{2} + k\pi$$

The second equation becomes

$$\sin\left(\frac{\pi}{2} + k\pi\right)\sinh\beta = \mp\frac{1}{\epsilon}$$

$$(-1)^k \sinh\beta = \mp\frac{1}{\epsilon}$$

$$\sinh\beta = \pm\frac{(-1)^{k+1}}{\epsilon}$$

This result determines a $\beta_0 > 0$ and a $\beta_1 < 0$. Thus we have

$$\arccos x = \frac{1}{n}[\alpha + i\beta_m]$$

$$= \frac{1}{n}\left[\frac{\pi}{2} + k\pi + i\beta_m\right] \qquad (m = 0, 1)$$

$$x = \cos\left[\frac{\pi}{2n} + \frac{k\pi}{n} + \frac{i\beta_m}{n}\right]$$

$$= \cos\frac{\pi(2k + 1)}{2n}\cosh\frac{\beta_m}{n} - i\sin\frac{\pi(2k + 1)}{2n}\sinh\frac{\beta_m}{n}$$

and we have our zeros. There are two sequences—$m = 0,1$ and $k = 0,1,\ldots,$ $2n - 1$—but only half actually appear (since $\beta_1 = -\beta_0$ and $\sin[\pi(2k + 1)/2n]$ has opposite signs as k comes down from $2n$ or goes up from zero). Thus the search can be confined to a single m, say $m = 0$.

The next step is to select the zeros in the upper half-plane. With our choice of m, the second term has a positive sign, and we confine the k values to 0, $1,\ldots,(n-1)$ so as to make the zeros fall in the upper half-plane.

The zeros must be paired off, from symmetric positions about the middle, and the corresponding real quadratics formed. On finally returning to the z variable, we find that we have a product of second-order terms, numerator and denominator, plus one possible first-order term. Again we identify the coefficients with the corresponding second-order digital filter and have the cascade of filters to use in obtaining the desired, stable transfer function.

12.6   CHEBYSHEV FILTERS, TYPE 2

The type 2 Chebyshev filters have the ripples in the stop band and are smooth in the pass band (Fig. 12.6-1). We start with the function (again in the w variable)

$$\tilde{H}(f)\tilde{H}(-f) = \frac{1}{1 + \epsilon^2\left[\dfrac{T_n(w_s/w_p)}{T_n(w_s/w)}\right]^2}$$

At $w = w_p$ the condition

$$\tilde{H}(f)\tilde{H}(-f) = \frac{1}{1 + \epsilon^2}$$

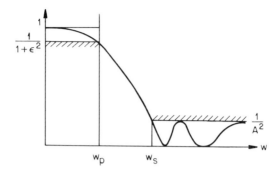

FIGURE 12.6-1   CHEBYSHEV TYPE 2 FILTER

is automatically satisfied. At the edge of the stop band we have $w = w_s$, and

$$\frac{1}{A^2} = \frac{1}{1 + \epsilon^2 T_n^2(w_s/w_p)}$$

Take the reciprocal of both sides

$$A^2 = 1 + \epsilon^2 T_n^2\left(\frac{w_s}{w_p}\right)$$

Finally, we solve for n, the design parameter,

$$n = \frac{\operatorname{arccosh}\left[\sqrt{A^2 - 1}/\epsilon\right]}{\operatorname{arccosh}\left[w_s/w_p\right]}$$

which is the same formula as for type 1 Chebyshev filters.

We next find the zeros of the denominator, take those in the upper half-plane, pair them off properly, select the second-order sections, go back to the z variable, and identify the corresponding coefficients of the second-order filters (with possibly an additional first-order filter). Again the cascade of the filters gives the transfer function sought. The details are as messy as those of the last section and are hardly more enlightening when displayed in full. Thus we omit them here.

12.7   ELLIPTIC FILTERS

Elliptic filters have ripples in both the pass and stop bands, but their design, as the name implies, depends on the theory of elliptic functions and hardly belongs in an elementary course. So, regretfully, we omit them with the observation that if the reader wants such a filter, there are advanced books in which they are described in some detail.

It should be evident, however, that by placing zeros in the numerator in the pass band (spaced in a Chebyshev spacing fashion with the appropriate density) and zeros in the denominator in the stop band, (again in a Chebyshev spacing and appropriate density), the result will be a filter that approaches equal ripple in the pass and stop bands. A few iterations (see Sections 9.7 and 12.8) of placing the zeros should result in a suitable design for that particular choice of the number of zeros in the numerator and denominator. It may be worthwhile to try a few different choices if the filter is critical for further work.

12.8 LEVELING AN ERROR CURVE

In applying a Chebyshev minimax criterion to a problem, we often find that we have an approximation whose error curve is close to, but not exactly, equal ripple. In such situations, if we want an exactly equal ripple error curve, we can use the error to find a new approximation, thereby using a feedback loop to find a level error curve ("level" means that all the local extreme errors are of the same size, not that all the errors are the same). Various methods are used, depending on personal preference. Some believe that it is merely a matter of the process of optimization—a question not of the result but of how to get there. We take the view that machine time for the average person is not important (although it may be for the specialist filter designer), so that almost any plausible optimization method is useful. Library optimizers tend to have trouble with the Chebyshev criterion, since the surface of the quality-of-fit function is apt to have corners.

One of the most popular methods is to compute the position and amounts of each extreme value and then perturb each parameter a small amount, one at a time, noting the change at each extreme value in the error curve due to the perturbation. Next, a set of simultaneous linear equations is set up to make the errors equal sized, and the equations are solved for the appropriate changes in the parameters. By repeating this process a number of times, we can get arbitrarily close to the equal level solution.

Another method is to note the approximate locations of the zeros of the error curve, moving the zeros together to reduce the larger extremes in the error curve and moving them apart so as to increase the smaller extremes. Again a few iterations will usually level the curve sufficiently for all practical purposes.

It should be noted, however, that almost any good optimization program will be capable of leveling the curve, *provided* that the proper criteria are fed into it. For problems in which the parameters enter linearly, the approach to the best solution is generally both rapid and reliable; but when the parameters enter nonlinearly, there is no guarantee that the general case has a unique minimum—instead there may be several local minima of the maximum error. All that can be done is to find a good, plausible starting value for the first trial and hope that it is in the right valley. Using various starting functions will give an idea of the size of the local valley, for if the functions all come down to the same minimum, they are in the same valley (probably).

12.9   A CHEBYSHEV IDENTITY

In order to carry out the design of an integrator in Section 12.10, we need the identity

$$e^{izs} = J_0(z) + 2 \sum_{n=1}^{\infty} i^n J_n(z) T_n(s)$$

where the coefficients $J_n(z)$ in the expansion in Chebyshev polynomials are the Bessel function. Since this identity is rarely given in the literature, we shall derive it along with a few properties of the Bessel functions.

We begin with the definition of the Bessel functions by means of the "generating function"

$$e^{(z/2)(t - 1/t)} = \sum_{n=-\infty}^{\infty} J_n(z) t^n$$

The substitution $t = -1/t'$ leaves the left side unchanged; and since the expansion is unique, it follows that

$$J_n(z) = (-1)^n J_{-n}(z)$$

or, as we need it,

$$(i)^{-n} J_{-n}(z) = i^n J_n(z)$$

For $z = 0$ we have

$$e^0 = 1 = \sum_{n=-\infty}^{\infty} J_n(0) t^n$$

and conclude that

$$J_0(0) = 1 \qquad J_n(0) = 0 \quad (n \neq 0)$$

From the form

$$e^{(z/2)(t-1/t)} = e^{z(t/2)(1-1/t^2)}$$

it is easy to see that the Bessel functions are even or odd, depending on whether the index n is even or odd.

Expanding the exponential of the generating function

$$e^{z/2(t-1/t)} = \sum_{n=0}^{\infty} \frac{[(z/2)(t-1/t)]^n}{n!}$$

and arranging in powers of t, we see that

$$J_n(z) = \frac{1}{n!}\left[\frac{z}{2}\right]^n + \text{higher powers in } z$$

Set

$$t = ie^{i\theta}$$

Then

$$z\left[\frac{t-1/t}{2}\right] = zi\left[\frac{e^{i\theta} + e^{-i\theta}}{2}\right] = iz\cos\theta$$

and

$$e^{iz\cos\theta} = \sum_{n=-\infty}^{\infty} i^n J_n(z)e^{in\theta}$$

$$= J_0(z) + 2\sum_{n=1}^{\infty} i^n J_n(z)\cos n\theta$$

Finally, we set $\cos\theta = s$ in order to get

$$e^{izs} = J_0(z) + 2\sum_{n=1}^{\infty} i^n J_n(z)T_n(s)$$

12.10   AN EXAMPLE OF THE DESIGN OF AN INTEGRATOR

Here we apply the identity in Section 12.9 to a specific example, Tick's integration formula of Section 3.4. It should be noted that Tick found the formula by computing the transfer function (the frequency response) of a family of formulas, searching until he found the one that he wanted. This method (given in Section 12.8) should not be regarded as being "too simple to use." We are merely using the problem to illustrate a design method.

The function that Tick wanted was the integral $y_n(t)$ of the measured data $y'_n(t)$. He imposed two requirements on the general form of a recursive filter

$$y_{n+1} = y_{n-1} + ay'_{n+1} + by'_n + ay'_{n-1}$$

Clearly, the y'_n are the usual x_n, the integrand values. First, he required that $y'(t) = c$ be integrated exactly. This means

$$2 = 2a + b$$

Secondly, he wanted the error curve at any frequency to be Chebyshev in the lower half of the Nyquist interval. His symmetric form meant that he had no phase errors in this recursive filter.

Suppose, as usual, that we have

$$y(t) = e^{i\omega t} = e^{2\pi i f t}$$

and are sampling at unit spacing. In order to get an expansion in Chebyshev polynomials, we use the identity in Section 10.9. To do so, we need to identify the variables z and s in

$$e^{2\pi i f t} = e^{izs}$$

Let the Chebyshev condition apply to the fraction λ of the Nyquist interval (we will put $\lambda = \frac{1}{2}$, finally, to get Tick's formula). Since $-1 \le s \le 1$ corresponds to $-\lambda/2 \le f \le \lambda/2$, we must have

$$s = \frac{2f}{\lambda}$$

$$z = \pi \lambda t$$

and therefore

$$e^{2\pi i f t} = J_0(\pi \lambda t) + 2 \sum_{n=1}^{\infty} i^n(\pi \lambda t) T_n\left(\frac{2f}{\lambda}\right)$$

Substituting it into the integration formula, we obtain

$$J_0(\pi\lambda) + 2\sum_{n=1}^{\infty} i^n J_n(\pi\lambda)T_n(s) = J_0(-\pi\lambda) + 2\sum_{n=1}^{\infty} i^n J_n(-\pi\lambda)T_n(s)$$
$$+ a\pi\lambda\left[J_0'(\pi\lambda) + 2\sum_{n=1}^{\infty} i^n J_n'(\pi\lambda)T_n(s)\right]$$
$$+ b\pi\lambda\left[J_0'(0) + 2\sum_{n=1}^{\infty} i^n J_n'(0)T_n(s)\right]$$
$$+ a\pi\lambda\left[J_0'(-\pi\lambda) + 2\sum_{n=1}^{\infty} i^n J_n'(-\pi\lambda)T_n(s)\right]$$

Because of the oddness and evenness of the Bessel functions, the coefficients of $T_{2k}(s)$ all vanish.

The coefficient of $T_1(s)$ will be zero if (dropping a $2i$ factor)

$$J_1(\pi\lambda) = J_1(-\pi\lambda) + \pi\lambda[aJ_1'(\pi\lambda) + bJ_1'(0) + aJ_1'(-\pi\lambda)]$$

or
$$2\pi\lambda aJ_1'(\pi\lambda) + \pi\lambda bJ_1'(0) = 2J_1(\pi\lambda)$$

Using $J_1'(0) = \frac{1}{2}$ and the earlier condition

$$2a + b = 2$$

gives

$$a = \frac{2J_1(\pi\lambda) - \pi\lambda}{\pi\lambda[2J_1'(\pi\lambda) - 1]}$$

Table 12.10-1 is easily computed.

Clearly, $\lambda = 0$ is Simpson's formula. This table provides a convenient set of integration formulas whose maximum error in the lowest λth fraction of the Nyquist interval is minimal.

TABLE 12.10-1

λ	a	b
0.0	0.33333	1.33333
0.1	0.33425	1.33150
0.2	0.33703	1.32594
0.3	0.34177	1.31647
0.4	0.34862	1.30276
0.5	0.35785	1.28431
0.6	0.36979	1.26043
0.7	0.38493	1.23014
0.8	0.40394	1.19213
0.9	0.42771	1.14457
1.0	0.45752	1.08496

The result at $\lambda = \frac{1}{2}$ is very close to Tick's result ($a = 0.3584$) and differs because our error term looks like $T_3(s)$, *but* it includes higher terms $T_5(s)$, and so on, in the Chebyshev expansion of the error of fit of the two sides, whereas his was carefully leveled by repeated computation of the error curve.

This discussion illustrates one method of designing integrators. Many others are readily devised. Clearly, recursive filters are necessary to "remember" the initial conditions in problems like integration.

13

Some Practical Details

13.1 TYPES OF FILTER DESIGN PROBLEMS

Filtering problems can be conveniently divided into two extreme types. At one end are those that occur in telephony, radio, and television systems. Here the design criteria are mainly determined by the systems design, and the filter designer rarely sees the filter in action. He thinks not of a particular set of data but rather of an ensemble of signals, all having cerain properties in common—primarily bandwidth restrictions.

At the other extreme is the person facing a single, unique set of data that he wishes to understand. For instance, suppose that the information consists of the number of bank failures per year for each of the past 50 years. What is wanted from the data? Understanding, of course! But what are the ways of looking at the data? The analyzer of the data will soon think of the probability of bank failure as a function of time. Further thought may bring him to think (or he may decide it is inappropriate) about the probability $p(t)$ of a single bank failure as a function of time. Thus the number of failures per year should (perhaps) be normalized by the number of banks in existence at that time. He next remembers that he has only a particular realization of this process of bank failures, one outcome of the function $p(t)$ acting on the banks at time t. Actually, he does not have this information; rather he has the data grouped by years. Therefore the original data that he is imagining has been passed through a Lanczos window of length one year and then sampled at yearly intervals. Finally, the series has been truncated to 50 years.

Earlier we carefully discussed all these operations on data, and the reader should now be well aware of the effects that they have on the spectrum of the data.

Often data is gathered and analyzed in order to make extrapolations to the future. If the unwary reader uses the typical polynomial extrapolation formulas, he immediately condemns himself to "extremist predictions," since it is the nature of a polynomial to rush off to plus or minus infinity as soon as it is released from the restraints of the data. On the other hand, it is often reasonable to try to understand the underlying mechanisms a little before rashly extrapolating. The spectral analysis approach is valid only if the system is linear, or close to linear, in its effects. If spectral analysis does seem to be the proper step, then the analyzer looks at a spectrum of the data, properly windowed if necessary, of course.

At this point we part from an introduction to digital filtering and enter the fields of statistics and economics, thereby dropping the subject of data analysis.

Most cases lie between these two extremes of (a) practically infinite runs of data whose characteristics are determined greatly by the systems design and (b) single runs of data, usually very short and not practically extendible. The analysis of data is clearly a branch of statistics. This textbook has been primarily concerned with building filters and has avoided the subject of which filters to select for use in the analysis. Thus the reader who must carry out some data analysis is advised to consider carefully the contributions that statistics can make and not assume that anyone (meaning yourself) can analyze data without any special statistical tarining. The topic is an old one, and there is a vast background of experience that can help the beginner. See [B1].

13.2 FINITE ARITHMETIC EFFECTS

We have discussed the computations to be done by the filter as if the arithmetic would be handled accurately and without severe roundoff. In practice, there are times when the finite word length of the numbers being used can have severe effects.

The beginner is apt to feel that although chance roundoff could occasionally affect the arithmetic, if the data is at all numerous, then he would be protected by the central limit theorems of probability, and only a very bad break would do him harm. He overlooks the fact, however, that when a coefficient of a filter is rounded off, this same roundoff occurs *every* time the coefficient is used! It is for this reason, among others, that we have recommended that the *final* filter design have its transfer function computed and drawn to see if anything peculiar has happend—to see if, because of roundoff in the coefficients, the actual filter will, in practice, differ significantly from the proposed filter.

To illustrate the strange things that can occur, consider the very simple recursive filter

$$y_n = 0.04x_n + 0.96y_{n-1}$$

with $x_m = 100$ for all m. Suppose that we start with $y_0 = 85$. We get the table [B, p. 76]

x_k	Computed y_k	Rounded y_k
100	–	85
100	85.60	86
100	86.56	87
100	87.52	88
100	88.48	88

and see that we are "stuck" at $y_n = 88$. The mathematical solution is, of course, for n approaching infinity

$$y_n = 100$$

If we started at $y_0 = 115$, we get

x_k	Computed y_k	Rounded y_k
100	–	115
100	114.40	114
100	113.44	113
100	112.48	112
100	111.52	112

and we are "stuck" at 112. Thus there is a "dead band" of width 24 units in this recursive filter.

Another effect that the reader is likely to overlook consists of limit cycles. A steady input may force a recursive filter to go through a cycle of values. Overload is also a problem. Exactly what will happen when finite arithmetic is used?

All such questions become increasingly important when the filter is to be built out of definite hardware—typically, integrated circuit chips—and used in the field (instead of software programs on general-purpose computers). Unfortunately, the chip is apt to have a rather short word length (to save money and perhaps the time needed to process the numbers). In this situation, a large gap between the simulation on a computer and the actual field processing of similar data is possible. The reader is referred to advanced books, reference [IEEE 1], and the current literature, for research is still active in this field.

13.3   RECURSIVE VERSUS NONRECURSIVE FILTERS

We have concentrated primarily on nonrecursive filters and have discussed recursive filters in only two chapters. When should one or the other be used? As noted, recursive filters can have a very short transition band for a comparatively short span of filter. But this statement does not mean that, given a short run of data, a recursive filter is required. The transient of the filter must also be considered: how long does it take for the filter to settle down after a sudden change? Fortunately, in any particular situation, the process can be simulated by a simple computation, and no great amount of theory is necessary in this very practical approach.

From a theoretical point of view, the important factor is the nearness of the zeros in the complex plane to the real frequencies (in the design-stage notation, the closeness of the zeros of the denominator to the unit circle in the z plane). If the transient of the filter takes a long time to settle down, what about properly windowing the input? About the same recommendations can be made as for computing the spectrum of data. By removing the mean (and possibly trends) and by using weights that are a cosine smoothing window, the transient can be somewhat minimized. The actual width of the cosine window is determined by the settling time of the filter (which, as noted above, can be found experimentally by running a step function into the filter and simply looking at the output).

Because of these problems, as well as others like phase shifts, recursive filters tend to be used in systems where there are very long runs of data; nonrecursive filters, which are simpler to understand, design, and use (for example, no instabilities to worry about) are likely to be used in most data processing situations where machine time is not a serious problem. Recursive filters also tend to have much less delay and are therefore used in real-time signal processing, such as telephony. Only occasionally is a recursive filter favored on the grounds that it uses less arithmetic for a sharp transition band. However, it is worth noting that both types of filters have about the same flexibility to meet various conditions when each has the same number of adjustable coefficients. It is only in one department, sharp transition bands that the recursive filter regularly shines. Of course, special conditions can be found in which any given filter is best; we need only select its transfer function and make this the design criterion to have it win the contest.

13.4 SPECTRAL ESTIMATION

Spectral estimation was referred to briefly in Section 10.7. Again, it is part of statistics and, as such, inappropriate for an introduction to digital filters. But we should clarify why noise frequently appears in the spectrum as a fairly flat contribution to the spectrum of the signal.

The spectrum of noise tends to be flat because the noise in the original analog signal *before* sampling often has a slowly decaying value as the frequency increases. It must fall off rapidly enough so that the noise has a finite energy, but often the falloff is slow. When the sampling occurs, the aliasing (see Fig. 10.3-1) *folds* the noise spectrum back and forth many times, and, as a result, the sum is fairly flat. (Remember that we assume that the noise is independent of the signal and of its own various parts.)

The noise that occurs from arithmetic computations, and from the quantization of the analog-to-digital converter, tends to be flat across the Nyquist band. Thus again we have a flat noise spectrum.

But not all noise spectra are flat. In practice, there are often special reasons for other shapes to occur, and the reader is advised to look carefully at the noise of his system before assuming that it has a flat spectrum. Although it is possible to do so in several ways, frequently, as in the example given in Section 13.1 of the number of bank failures, he can only study the sources of errors in the data and wonder how much they have contaminated his data. Before believing data too much, the careful reader should look at Morgenstern's book, *On The Accuracy of Economic Observations* [M].

Suppose that we have a slowly falling spectrum as our input signal. If, as is customary, the mean of the data is removed, then immediately the spectrum has a discontinuity at $f = 0$. If additional smoothing is done, say a von Hann window to reduce oscillations in the spectrum, the effect of the removal of the mean will be spread out a little. The value of the spectrum at $f = 0$ will be raised a bit from its zero value, the next one or two points will be corre-

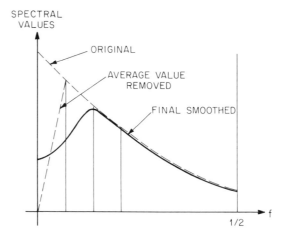

FIGURE 13.4-1   A FALSE SPECTRAL PEAK

spondingly lowered, and the resulting spectrum will show a suggestion of a real peak (Fig. 13.4-1). It is necessary to understand what has been done to the data before attempting to interpret the results that have been computed!

13.5   DECIMATION

Technically, "decimation" means taking every tenth, but, in practice, it usually means taking every other one of a set. This process occurs in many problems of data processing and so a few words of warning are necessary.

It is probable that the basic reason for decimation is that in real life we do not want equally spaced values in the spectrum; instead we prefer the frequency values much more densely spaced near the zero frequency. An almost logarithmic spacing in the frequency of the spectrum is often desired. As a result, the experimenter will take a very long run of data at a fairly close spacing. The close spacing is necessary in order to obtain the high-frequency components of the spectrum and to avoid the aliasing that would occur if a lower sampling rate were used. The long run of data is taken in the correct belief that low frequencies will require a large number of complete periods if they are to be accurately determined.

In going from the high rate of sampling to a lower one, we first filter the data to remove the upper half of the spectrum and then subsample by taking every other data point. This procedure avoids the aliasing that would arise from the subsampling. Frequent use of this process will reduce the volume of data to be run through the final spectrum computation to manageable amounts. The effects of the aliasing at each stage can more or less be compensated for by noting how far the filter is from ideal and multiplying the values in the upper part of the spectrum by the reciprocal of the error in the lower half of the previous spectrum.

However, note that in the early stages of decimation it is only necessary to be careful that no aliasing falls into the final band that is to be used. If other aliasing occurs, later stages of decimation will remove such frequencies.

Achieving a maximally efficient reduction is a complex problem, and one that is not suitable for this text [CR].

13.6 REFERENCES

References are a somewhat difficult problem in an elementary book that covers a rapidly developing field of research. By the time that the reader needs the references, they may be out of date. Furthermore, the beginner is not likely to be interested in advanced references. Consequently, we have given comparatively few references in the text.

The reference [IEEE-1], *Literature in Digital Signal Processing: Author and Permuted Title Index, 1975,* is very complete as of its date of publication, having been computer prepared and typeset. Thus the reader is referred to it as a general source of references.

Two other references, [IEEE-DSP1] and [IEEE-DSP2], are collections of published papers that were regarded as of special importance.

Presumably, further editions of the IEEE publications will occur and in time will provide the necessary up-to-date references.

The books by Oppenheim and Shafer [OS] and by Rabiner and Gold [RG] are excellent ones that start from the assumption that the analog filter theory is already known. The first is suitable for advanced undergraduates; the second is a more-advanced text. Stearn's book [S] is especially good on the sampling theorem, and is generally more mathematically oriented than the other two just mentioned.

REFERENCES

[B] BLACKMAN, R. B., *Data Smoothing and Processing.* Reading, Mass.: Addison-Wesley, 1965.

[B1] BLOOMFIELD, P., *Fourier Analysis of Time Series: An Introduction.* New York: Wiley, 1976.

[BT]   BLACKMAN, R. B., and J. W. TUKEY, *The Measurement of Power Spectra*. New York: Dover, 1958.

[Br]   BRIGHAM, E. O., *The Fast Fourier Transform*. Englewood Cliffs, N.J.: Prenctice-Hall, 1974.

[CR]   CROCHIERE, R. E., and L. R. RABINER, "Optimum FIR Filter Implementations for Decimation, Interpolation, and Narrow-Band Filtering," *IEEE Trans. Acoustics, Speech, and Signal Processing*, October 1975.

[H]   HAMMING, R. W., *Numerical Methods for Scientists and Engineers*. New York: McGraw-Hill, 1972.

[IEEE-I]   *Literature in Digital Signal Processing: Author and Permuted Title Index*. Revised and expanded edition, edited by H. D. Helms, J. F. Kaiser, and L. R. Rabiner. New York: IEEE Press, 1975.

[IEEE-DSP1]   *Digital Signal Processing*, edited by L. R. Rabiner and C. M. Rader. New York: IEEE Press, 1972.

[IEEE-DSP2]   Signal Proc. Comm. New York: IEEE Press, 1975.

[K]   KUO, F. F., and J. F. KAISER, *System Analysis by Digital Computer*. New York: Wiley, 1966.

[KS3]   KENDALL, M. G., and A. STUART, *The Advanced Theory of Statistics*, vol. 3. New York: Hafner, 1968.

[L]   LANCZOS, C., *Applied Analysis*. Englewood Cliffs, N.J.: Prenctice-Hall, 1956.

[Li]   LIGHTHILL, M. J., *Fourier Analysis and Generalized Functions*. New York: Cambridge University Press, 1960.

[M]   MORGENSTERN, O., *On the Accuracy of Economic Observations*, 2nd edition. Princeton, N.J.: Princeton University Press, 1963.

[OS]   OPPENHEIM, A. V., and R. W. SCHAFER, *Digital Signal Processing*. Englewood Cliffs, N.J.: Prentice-Hall, 1975.

[P]   PAPOULIS, A., *The Fourier Integral and its Appli-*
              cations. New York: McGraw-Hill, 1962.

[RG]   RABINER, L. R., and B. GOLD, *Theory and Appli-*
              cation of Digital Signal Processing. Englewood
              Cliffs, N.J.: Prentice-Hall, 1975.

[S]   STEARNS, S. D., *Digital Signal Processing.* New
              York: Hayden Book Co., 1975.

[T]   TUKEY, J. W., *Exploratory Data Analysis.* Read-
              ing, Mass.: Addison-Wesley, 1976.

Index

Digital filters *(cont.):*
  another design method, 178
  bandpass design method, 157
  Butterworth, 189
  causal, 3
  Chebyshev:
    type 1, 201
    type 2, 205
  definition of, 2
  elliptic, 206
  FFT design method, 177
  FIR, 181
  IIR, 181
  low noise, 107
  noise amplification, 13
  nonrecursive, 2
  physically realizable, 3, 181
  recursive, 2
  smooth, 118
    design of, 128
  smooth bandpass, 130
  testing, 109
  time invariant, 3
  two way, 194
Distribution of a statistic, 11

E

Eigenfunction, 15, 19
  of calculus, 24
  equally spaced, 25
  linear systems, 23
  translation, 22
Eigenvalue, 15, 20
Eigenvector, 20
Elliptic filter, 206
End effects, 3
Ensemble, 7
Equally spaced samples, 15
Euler identities, 22, 77
Expected value, 8

F

Fejer smoothing, 77
FFT, 174
Filter sharpening, 112
Finite Fourier series, 169
Finite sample size, 145
FIR filters, 181

Fourier coefficients, 60
Fourier integral, 132, 137
  inversion formulas, 133
Fourier series, 55
  and least squares, 64
Fourier shifting theorem, 140
Frequency, 15, 16
  angular, 16
  fundamental, 16
  rotational, 16

G

Gaussian (normal) distribution, 7, 12
  mean of, 9
  variance of, 10
Gibbs phenomenon, 73, 77, 85, 154
Gold, B., 219, 221
Grouping of data, 4, 163, 213

H

Hamming, R.W., 220
Hamming window, 90, 94, 98
Highpass filter, 96
How a filter works, 53

I

IEEE, 220
IIR filter, 181
Integration:
  midpoint rule, 39
  Simpson's formula, 39
  Tick's rule, 40
    derivation of, 209
  trapezoid rule, 37
Interpolation of midpoints, 118
Invariance:
  linear systems, 23
  translational, 22

K

Kaiser, J.F., xii, 106, 149, 153, 220
Kendall, M.G., 220
Kernighan, B.W., xii
Kuo, F.F., 220